T0244603

The Sexual Life of Flowers

The Sexual Life of Flowers

Simon Klein

CONTENTS

INTRODUCTION

Did you know that flowers make love too? More specifically, they pave the way for plants to do so. But how? Well, flowers are plants' sexual organs. No false modesty here, though. Molière's 'Cover that bosom that I must not see' is not a concept that goes down well with flowers, nor any other scruples uttered by more prudish souls, for that matter. Everything is there on show, for all to see – tempting the eyes and noses of any creatures passing by. Exhibitionism is the norm amongst our plant cousins.

Inability to move: therein lies the problem

There's a reason why flowers are so visible. And as we shall discover, this is primarily down to one very simple property associated with plants: they are unable to move. They have their roots in the soil and their stalk points the rest of the plant skywards. Depending on the season, this stalk may bear flowers. However, other than the wind blowing the branches and leaves now and again, plants remain rooted to the ground. And yet, in any other species on this planet, the act of reproduction requires movement: the sperm or male sex cells (also known as male gametes) need to unite with the ovules, or female gametes.

In humans, the ovules remain in the woman's body, while the man provides the motile sperm cells as part of the sexual act. This requires both the man and the woman to come together, after which the male gametes are discharged and fuse with the female gametes. The same process takes place in all mammals and, apart from a few exceptions, many species in the animal kingdom use very similar mechanisms.

A profusion of colourful spring blooms in Monet's garden at Giverny. Or plants flaunting their sexual organs?

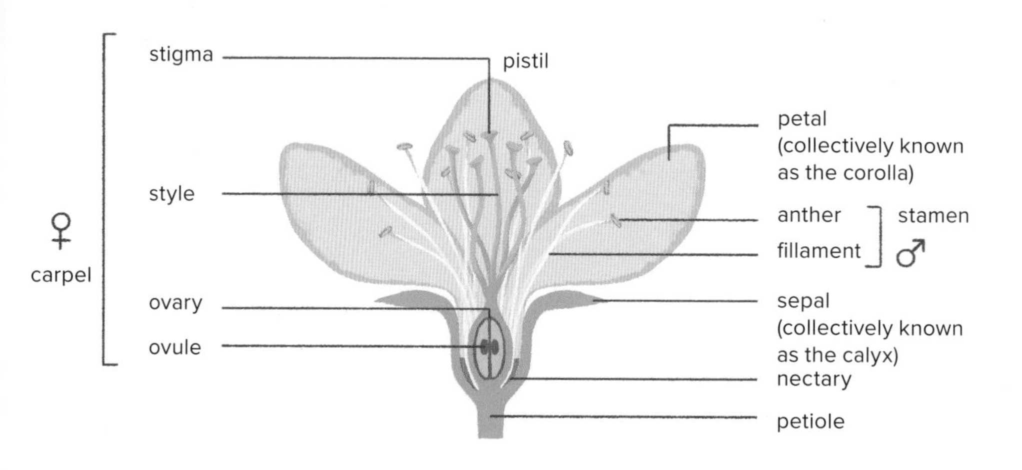
stigma
pistil
petal
(collectively known
as the corolla)
style
anther
stamen
fillament
♀
carpel
♂
ovary
sepal
(collectively known
as the calyx)
ovule
nectary
petiole

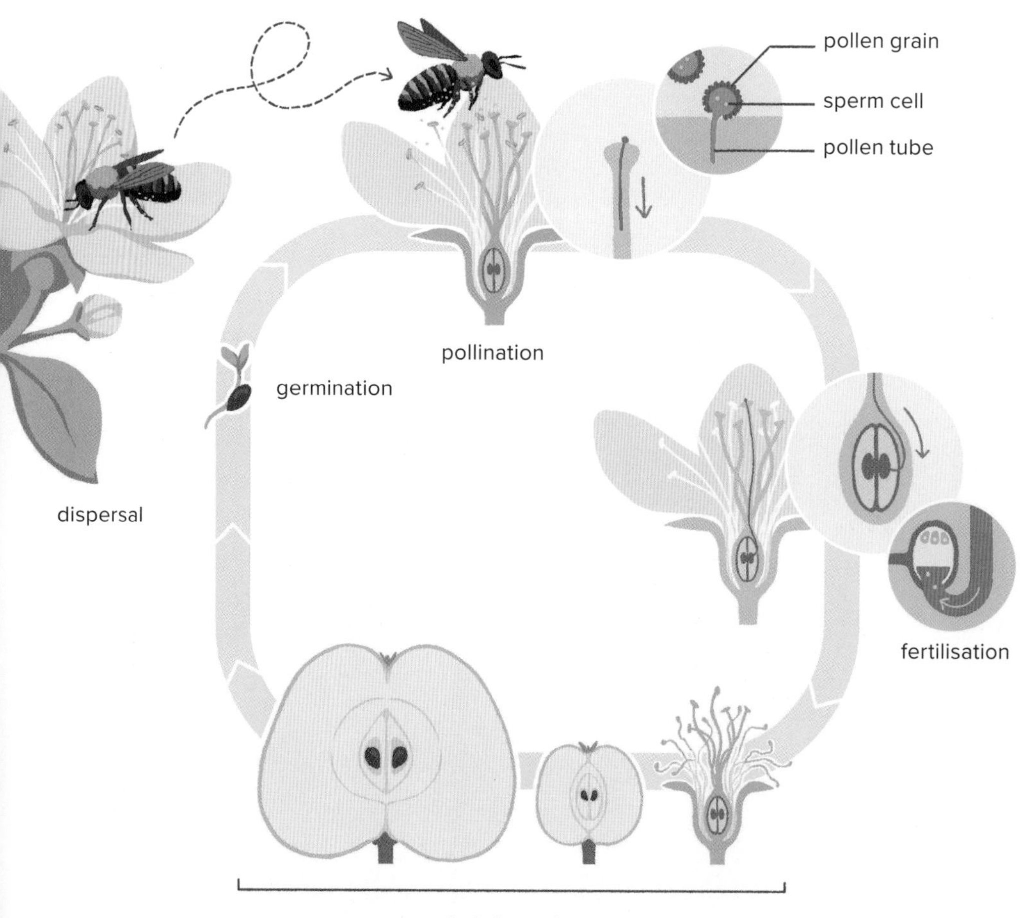
pollen grain
sperm cell
pollen tube
pollination
germination
dispersal
fertilisation
fruit formation

The sexual anatomy of plants

Let's take a closer look at flowering plants: sperm cells and ovules have a part to play here too. The spermatozoa, or sperm cells, are protected inside a small sphere, safe and sound within a microscopic grain of pollen, each one containing two sperm cells. All these tiny grains are closely packed together in a case-like structure known as the anther, which is in turn located right at the top of a fine stem called the filament. Together, the filament and the anther form what's known as the stamen. Therefore, by analogy, the stamen is the plant's male reproductive organ.

The ovule, on the other hand, is hidden deep in the heart of the flower, well protected inside the ovary, which is made up of plant tissue (a collection of cells) and will become the fruit once the flower has been fertilised.

Even though it is protected by the ovary, the ovule's main role is to be on standby to receive a sperm cell when the opportunity arises. The entrance to the ovary is via a complicated overhanging device in the form of a long corridor, or style, directed towards the sky. This is covered by a flat structure which is often bilobate, trilobate or even ribbed, known as the stigma. Like a door, the stigma provides the gateway between the world outside and the isolated world inside the flower, home to the precious ovule. The style and the stigma together form the pistil and, with the ovary, combine to make up the female reproductive organ, or carpel. In most flowers, the ovary is in the centre, surrounded by stamens, which are in turn encircled by petals to form the corolla, as well as another layer of plant matter called the calyx. This is often made of hard, green structures that protect the flower when in bud. This outer layer is made up of sepals.

The plant is said to be fertilised when the ovule combines with a sperm cell. They become fused together to form an egg cell that goes on to become an embryo and then a young plant in its own right, a blend of male and female genes.

Fertilisation requires a transport mechanism before it can take place. But since plants can't move, they have developed strategies to bring about this fusion. And it is precisely this evolutionary process that has led to the often extraordinary methods by which plants reproduce.

Pollination: its place in the long history of reproduction

Over the course of evolution, the arrival of flowers marked a turning point in the quest to circumvent the problem of immobility. As soon as pollen appeared on the scene, transportation became a possibility. In other words, the plant itself doesn't move, merely its tiniest units, pollen grains, which in turn transport the sperm cells. It is thought that the act of pollination dates back some

250 million years BC to the Triassic period. In plants, fertilisation occurs twice: one of the sperm cells comes into contact with the ovule, while the other comes into contact with the surrounding tissue in the part of the ovary known as the micropyle.

This double fertilisation process allows seeds and fruit to form. If this fruit is fleshy and surrounds the seed, it plays a very important role in reproduction as it attracts animals, which eat the fruit and swallow the seeds. The fruit is digested, but this rarely happens with the seed, which tends to be rather solid and can be found intact in animal faeces deposited hundreds of metres from where it was first eaten. This seed will in turn become a new plant. This is a success story for the species in question: once again, it has overcome the problem of immobility to colonise the surrounding environment and thrive within the ecosystem: this process is referred to as dispersal. If we examine plant reproduction in more detail, we need to bear in mind that it comprises a number of successive stages involving a range of strategies used by plants to compensate for their immobility. These stages include pollination in the first instance (during which the pollen is transported), immediately followed by fertilisation, which takes place inside the flower, after which the seed and fruit grow to maturity and finally the seeds are dispersed. This might come about via animals, as in the case of fleshy fruit, or by the wind, as happens with dandelions, or even by water, such as when coconuts are carried from island to island.

Three bees (*Apis mellifera*) returning to their colony after gathering nectar and purple pollen from phacelia flowers (Phacelia tanacetifolia).

This book will focus on the first part of the plant reproductive cycle: pollination, which precedes fertilisation.

We'll leave the sequel to this story for the time being; it only starts to unfurl when the flowers have faded...

Wind pollination

What better carrier could you wish for than the wind? As a force of nature, it covers all four corners of the world (apart from a few dark forests where it's just too hot for either wind or pollinators – in these dark nooks and crannies, flowering plants have to resort to self-fertilisation, as in the case of Holcoglossum amesianum, an orchid found in China). Wind is able to carry grains of pollen over huge distances; pollen from the Scots pine has been found in Spitsbergen, some 750 km north of the Russian coast.

Yet wind pollination also has its downsides: pollen grains are dispatched into the air as soon as the anthers ripen and release them. This means they can end up anywhere, including on flowers where they will serve no purpose. Wind is an inaccurate carrier, leading to major

losses. As a result, flowers that use the wind for pollination produce vast quantities of pollen grains and these are often very small and light. Some flowers have even discovered a trick for making the best possible use of updraughts: each grain of pine pollen is equipped with two air balloons to make it even lighter.

Pollination by animals

An epic love story that first began some 120 million years ago, this tale of a symbiotic relationship between flowering plants and pollinating insects, notably the honeybee, has woven its way through the succeeding millennia, yet it's more than just a passion, it's a history of co-evolution. These species evolved at the same time, developing extraordinary links which have often led to them becoming dependent on one another. The way these animals feed and flowering plants reproduce is an outstanding example of good practice.

Beware of self-pollination!

If we are to understand why it's so important to have access to reliable pollen carriers such as bees, we need to remember that flowers abhor self-fertilisation. This is what happens when pollen from a flower lands on the pistil of the same flower (the sperm cell from one flower comes into contact with the ovule of the same flower). Given that over 80% of plants have flowers with both male and female organs (so-called hermaphroditic flowers), there is a very high risk of this occurring and it does in fact sometimes come about. Some species, like peas, do self-pollinate and, as a result, their breeding lines, and the long-term survival of the species may be in jeopardy. In other words, self-fertilisation and self-pollination are not dissimilar to inbreeding in humans. By keeping themselves to themselves, there is a complete lack of genetic mixing.

The genes of both parents are identical; this means that there only needs to be a tiny adverse mutation in the parent plant and this will be passed on to the next generation, where it may remain in the population ad infinitum. Such mutations may include deformed fruit or the plant being unable to withstand cold. However, with cross-pollination (where pollen is transferred to another flower), there is the hope that at least one of the two parents won't display this mutation and that their offspring won't have it either. In other words, cross-pollination is the holy grail of all plant species because it guarantees genetic diversity, along with the ability to evolve and adapt to the environment in which they grow, especially when events like forest fires and deforestation can cause habitats to change considerably.

How do flowers attract pollinators?

To create the ideal conditions to allow animals to carry their pollen, flowers have become adept at advertising their wares, displaying remarkable complexity and efficiency thanks to the use of different colours, shapes and scents. These are just as pleasing to the eyes and noses of human bystanders, who have played an active role in selecting varieties to brighten up gardens and balconies, as well as producing beautiful bouquets of flowers. So, these are essentially the seduction tools used by flowers to attract pollinators. The petals act as flags that enable flowers to be identified from a distance against a predominantly green background. The size of the flower is also significant: the larger the flower, the easier it is to spot.

Flowers emit scent when their pollinators are most active. These scents are picked up by the insects' antennae and are often the first marker leading the way to the flowers. which in turn seem geared to encourage visitor loyalty, tempting the insects to return as often as possible.

Even before humans evolved on the planet, and certainly before they invented money, flowers had worked out how to set up a barter system. Most flowers reward their carriers by

The iris (*Iris germanica*) attracts pollinators by its scent and complex, colourful structure.

providing them with food. The first part of the bargain is the precious pollen, which allows fertilisation to take place. But pollen is also a very important food source for many pollinating insects. Many flowers, roses and poppies among them, go all out to advertise their pollen stocks, producing large quantities to attract insects.

A great many flowering plant species reward insects for their hard work by providing them with nectar. This is a sugary solution rich in vitamins and natural antibiotics, which is produced in nectaries at the base of the flower, below the stamens and the pistil. This nectar is collected by honeybees, dehydrated and transformed into the precious resource of honey, providing bees with valuable reserves to see them through the winter months. Honey provides the bees with energy to keep the hive warm. While pollen is an important source of proteins and lipids, which are particularly useful for larvae in their growth phase, nectar (and therefore honey) is a source of sugar used for motor functions such as flight or shivering – which the bees do to increase the temperature in the hive in the depths of winter.

This liquid is so beneficial for the various animals visiting the flowers that some insects like bumblebees have found a way to access this precious booty when it's hidden away deep inside the flower: they drill a little hole at the base of the flower and then suck out the nectar from the outside. But this constitutes theft and definitely isn't playing the pollination game – their bodies don't even come into contact with the pollen in this scenario.

How do they end up covered with pollen?

While the purpose of flowers is to flag up a food supply for certain animals, the architecture of the flowers is cleverly designed to make sure that any visitors leave smothered in pollen grains. Any pollen remaining after insects have finished grooming themselves, a very frequent occurrence, may possibly end up on the pistil of another flower. There's a great deal of wastage, however. To give you an example, only 3% of the pollen from Disa orchids reaches another plant. The rest remains in situ, where it may rot, be used for self-pollination, carried away by the wind or washed away by water – or may quite simply be eaten. Then there are potential losses in transit via the pollinator. All of which means that, in this particular orchid, only 1% of the pollen produced by one flower is likely to reach the pistil of another flower.

A huge range of pollinators

Despite this significant wastage, pollination by animals is still more efficient than wind pollination. So effective is it that the wider flower family has elected to diversify how it interacts with the animal kingdom: some flowers are pollinated by birds (such as the bird of paradise flower

A buff-tailed bumblebee (*Bombus terrestris*) caught in the act of stealing nectar from an abelia flower (*Abelia grandiflora*).

The centre of a poppy (*Papaver rhoeas*) is darker than the rest of the corolla so that it retains heat.

The arum lily (*Zantedeschia aethiopica*) attracts flies by emitting odours that are typical of their natural egg-laying sites such as animal droppings.

and hibiscus), others by bats (like callistemon, banana plants and the baobab tree), and, even more rarely, some are pollinated by lizards (such as Roussea), small marsupials (callistemon) or even snails (some species of morning glory). Insects such as honeybees are often quoted as examples of pollinators, but we should remember that 90% of bee species are neither domesticated nor social animals and that butterflies, beetles, flies or even ants also have a major role to play in pollination.

Life doesn't just revolve around food!

Although food is vital for all creatures, it isn't the only trick up a flower's sleeve. Flowers can also supply pollinators' other vital needs. Warmth is a case in point. Many flowers like poppies, tulips or sunflowers provide warmth for insects that tend to be on the chilly side... Other plants provide meeting places – with giant waterlilies playing host to tiny beetles – or even egg-laying sites, as in the case of fig trees and the fig wasp, Blastophaga psenes. Some flowers even go so far as to put out misleading advertising messages: they imitate egg-laying sites (arums), or even sexual partners, as in the case of certain orchids – anything

to attract those sought-after pollen carriers!

Lifting the veil on botany

These natural exchanges between flowers and animals, or even between flowers and environmental features such as wind or water, give rise to extraordinary stories that bear witness to the complex ways in which plants have evolved in order to survive. In this book, we have selected 50 of the most important examples – in our opinion at least. It goes without saying that we could have written a great deal more on this subject, if not several books' worth, if only because the mystery and magic of pollination is evident in a great many species, from the simple daisy right up to the grandest chestnut tree. But our aim, with this book, is to provide an introduction to this subject, an invitation to observe the living world before attempting to explore it further.

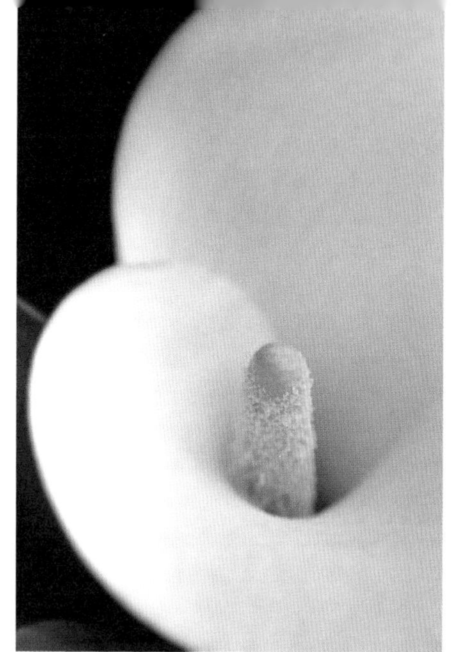

We hope to delight you with these tales of charm, seduction and intrigue, all playing out before our very eyes and creating an explosion of springtime colour in our meadows, woodlands, verges and fields while allowing biodiversity to flourish.

You may decide to use this book as a field guide, or perhaps even as a story book. However you use it, it will make you aware of these fantastic stories wherever and whenever you are out and about. Or you can simply leaf through it and wonder at Mother Nature's miracles.

The importance of pollination for humans

These plant species are vital if humans are to survive on Earth. Most of what we eat comes from plants that produce flowers or are even flowers themselves: you might well have eaten deep-fried courgette flowers or perhaps even elderflower fritters? But even if these delicacies have passed you by, we've all eaten vegetables, fruit or cereals, and used all kinds of flour or oil in cooking. Plants form a fundamental part of the human diet and all plants will have started off as seeds before being harvested. In other words, every plant is the result of an act of pollination. When you think about it, more than two-thirds of our food is linked to pollination by animals. The remaining third is mainly made up of cereal crops grown across the countryside (wheat, barley, corn, etc.), all of which are wind-pollinated. If pollinating insects were to disappear from the face of the earth, human food security would be at serious risk.

Who wants to live on a daily diet of wheat or corn when all's said and done? Ten years ago, scientists even put a price on pollination by animals: it was thought to be worth some €153 billion a year.

Threats to flower sexuality

The gradual disappearance of pollinating insects is already well underway... When we used to spend hours in the car to reach our summer holiday destinations back in the 1980s, it wasn't unusual for the windscreen to become so splattered with tiny dead insects that we had to make frequent stops to clean it. Yet nowadays this irksome summer chore has all but disappeared. This is tragic proof, if proof were needed, of the extraordinary decline in the number of insects over a period of just 30 years! This finding was confirmed by a Danish study that recorded an 80% drop in the number of insects crushed on windscreens between 1997 and 2017. Other more complex studies have found the same trend: a rapid reduction in very many types of flying insects. This trend is particularly noticeable in Europe and the United States. Why has this happened? The insects' habitat has been destroyed by agriculture and urbanisation, along with extensive use of pesticides, and climate change has caused certain species to migrate or has led to them becoming extinct. It is possible to assess the health or otherwise of honeybee populations, and the decline in colony numbers over many years, across a wide range of countries, is much greater than natural rates would suggest. These trends are undoubtedly linked to a large extent to the widespread use of pesticides and the scarcity of insect food sources, but also to increasing pressures from predators or parasites and to climate change itself.

Climate change has an insidious role to play in disrupting the harmonious relationship between flowers and pollinators: it changes flowering times. Flowers are starting to open earlier, while insects, which are less susceptible than plants to these changes, are only active a little later. This phenomenon is only too apparent in almond and cherry trees: it means the plants are unable to reproduce and insects no longer have access to their food source.

Human activities have a significant impact and the ability to disrupt the long and magnificent history of pollination. We hope you will find this story as fascinating and as stunning as we do.

Apple orchards in full bloom in springtime in the South Tyrol; these trees attract a great many insects, which are essential for pollination and hence for the production of fruit.

Next page (double spread)

The most important pollinators are from the Hymenoptera order, e.g. honeybees, buff-tailed bumblebees, European orchard bees (Osmia cornuta) or carpenter bees. Then you have the Diptera order: hoverflies, Psychoda flies; followed by Lepidoptera such as the peacock butterfly or hummingbird hawk-moth. Some members of the Coleoptera order, including ladybirds or beetles, are also pollinators. Other Hymenoptera species may occasionally pollinate flowers too; these include ants and wasps.

Tropical flowers are pollinated by a great many birds, notably hummingbirds. There is also anecdotal evidence to suggest that bats, lemurs, marsupials, geckos and even one particular species of small snail can pollinate specific flower species.

BINDWEED & IPOMOEA

Trumpets of glory

Gardeners have a love/hate relationship with these climbers. One is an elegant plant adored for its beautiful flowers, but the family also includes a despised plant to be hacked down at all costs, as it has a tendency to become invasive. Both belong to the same family of flowers with similar properties: Convolvulaceae. The name comes from the Latin verb *convolvulare* – to entwine – as they have a habit of winding themselves around any support. One of the main characteristics of these climbers is that they tend to suffocate any plant unfortunate enough to be used as a support, swamping them in the process. One member of the family (bindweed) can be found in the Old World, while the other, native to the Americas, has become naturalised in many gardens on account of its stunning purple flowers.

Convolvulaceae flowers are very distinctive and their anatomy guarantees successful cross-fertilisation. They have large funnel-shaped (or infundibular) flowers. This shape is practical in a number of respects: it protects the reproductive organs at the base of the flower and it provides a simple way of locating the nectar and pollen: right in the middle. This allows insects to approach the food source from any direction. In both bindweed and ipomoea (commonly known as morning glory), protecting the reproductive organs is a family trait: bindweed flowers close up whenever the sun goes in or at any sign of a cloud heralding rain that might discourage potential pollinators and risk damaging the heart of the flower. Ipomoeas, on the other hand, shut up shop at dusk to protect their flowers from nocturnal predators. As soon as day breaks, the corolla unfurls back into its typical infundibular shape; this is where its common

SCIENTIFIC NAME
Convulvus sp. and *Ipomea purpurea*

FAMILY
Convolvulaceae

HABITAT
Bindweed: fields, wasteland, roadside verges
Ipomoea: tropical regions, open position

WHERE TO SEE
Ipomoea can be grown from seed and cultivated in gardens. Bindweed can be found in its natural habitat.

FLOWERING SEASON
June-October

STRATAGEM

The funnel-like shape of these flowers makes it easier to access the nectar, located right in the heart of the flower, while forcing insects to pass over the reproductive organs. This means the pollinator gets covered in pollen or has to deposit pollen from another flower on the pistil.

name, morning glory, comes from.

The corolla is made up of five petals, joined from the base to the tip. This geometry continues inside the flower, with five pollen-bearing stamens, which emerge from five carpels that ensure access to the nectar. The pistil projects from the centre of the stamens, is slightly longer and therefore away from the pollen source inside its own flower: this is a good way of avoiding self-pollination.

Once a bee or bumblebee identifies the Convolvulaceae flower by sight (these flowers have no scent, so flower fragrance doesn't have any part to play here) and flies in, it plunges its head, and with it, its whole body, right down to the bottom of the flower, inserting its tongue into one of the carpels, where it proceeds to feast on the nectar. On quickly exhausting the nectar supply in that carpel, it inserts its tongue into the next and the process starts again. One by one, the Hymenoptera will systematically empty the five nectar-filled carpels, not unlike the chambers of a revolver barrel. This structure forces the bee to visit all the nectar-bearing carpels before it has had its fill, keeping it in the flower for longer, moving all the time. In the process, it becomes coated with pollen, which is why this kind of structure is often described as having a 'revolver architecture'. Once satiated, the bee flies off, its back smothered in pollen, only to spot another Convolvulaceae flower from the same species and fly straight in. (As long as the insect is among bindweed flowers, it will keep on going as it's easier than having to think – otherwise known as flower constancy.) As it enters, there's a good chance it will come into contact with the stigma borne by the long style of this flower, depositing the pollen gathered from the first flower. And there you go: the bee has played its part in the cross-fertilisation cycle!

From top to bottom

Ipomoeas close their flowers at night to protect the pistils and stamens.

Waking Up: A Girl of the Kōka Era, Tsukioka Yoshitoshu, 1888.

Blue rock bindweed (*Convolvulus sabatius*) has characteristic funnel-shaped flowers.

This kind of fertilisation is very typical among Convolvulaceae flowers and seems to have led to a very strange relationship observed in one exotic species, *Volvulopsis nummularium*, which is found in India. This member of the Convolvulaceae family is still infundibular, but does not shy away from rain as bindweed tends to; on the contrary, although it is pollinated by Asian honeybees on sunny days, it remains open even when it rains, in the hope of attracting a rather unusual visitor. On rainy days, a small snail pokes the tip of its shell deep inside the funnel-shaped flower of the *Volvulopsis*. *Allopeas gracile* (graceful awlsnail), with its pointed shell, is a frequent visitor and pollinator of *Volvulopsis*.

POPPY
Black hearts

Freshly ploughed soil is perfect for poppies and it's in springtime that they can be seen at their best: huge red flowers with black hearts spilling over field margins before the crops are ready to harvest. The poppy (or field poppy) is one of the cornfield flowers that tends to grow amongst grain crops. The history of the poppy is closely linked to the history of agriculture itself. The field poppy is the descendant of a plant which is still found to this day in the Middle East, in Mesopotamia, where agriculture first originated. As people began to travel further afield, agricultural practices spread throughout the world and the poppy evolved to accompany this surge in global agriculture.

The poppy flower is as graceful as it is fragile, and extremely short-lived. The hairy, green buds, shaped like miniature rugby balls, start to stand tall in the early morning, then the petals, creased and rumpled inside the bud, emerge from their protective casing and begin to unfurl. In no time at all, four brilliant red petals spread out like wings, losing their crumpled early-morning demeanour in the process. By 7 o'clock, the flowers will be fully open: let the feast begin! Right in the heart of this scarlet bowl is a veritable banquet: a black coronet comprising hundreds of stamens containing billions of grains of pollen ready for all comers. No nectar, though: the poppy is one of the very few flowers that only rewards its pollinators with pollen. And the poppy is dependent on insects; the anatomy of the flower is not designed for optimum anemophily (wind pollination). They rely on pollen from another flower if they are to form seeds: the rows of stigmas arranged in a star shape around the large central style

SCIENTIFIC NAME
Papaver rhoeas

FAMILY
Papaveraceae

HABITAT
Cereal crop and cornfield margins, banks, roadside verges

WHERE TO SEE
Just before the crop starts to turn golden, on roadside verges, often accompanied by cornflowers. Some species of horticultural Papaver can also be cultivated in gardens and on balconies. The opium poppy is a Papaver grown on a large scale in fields to produce drugs such as morphine.

FLOWERING SEASON
May-September

STRATAGEM

The flower is red, a colour that doesn't attract bees but emits ultraviolet light and signals the presence of pollen at its heart by a difference in heat that insects pick up on. Ultraviolet light and heat, neither of which is visible to the human eye, are very useful in helping insects track down pollen, which the poppy produces in vast quantities.

will only accept pollen from other flowers.

The reason why poppies manufacture so much pollen is that a large proportion of these grains will end up in the pollen baskets on the hind legs of bees and bumblebees after they wallow in this profusion of pollen and take it back to their colony. But this free-for-all is very short-lived; by 11 o'clock the scarlet petals will start to droop and then flutter to the ground by midday. The truth is that poppy flowers usually only last for a morning, or perhaps one day if they're lucky.

But over the course of one morning the fabulous red flower will have lured in hordes of insects, especially honeybees and bumblebees. Don't forget, though, that these insects find it very hard to pick out red flowers as they don't have any specific receptors for this colour. In their eyes, red comes across as a greyish shade, difficult to distinguish against the green backdrop. However, these insects are sensitive to ultraviolet light, which is invisible to the human eye. Poppy flowers, red as far as humans are concerned, emit ultraviolet rays that the insects pick up on – and this is how they stand out against their surroundings. To further enhance visibility and lure in pollinators with even more precision, the black markings at the base of the corolla act as beacons to flag up the pollen source. In other flowers, these serve as nectar guides.

Interestingly, the Mesopotamian species from which our national poppy originates is predominantly pollinated by beetles, which are very good at discerning the colour red. The petals of this Middle-Eastern poppy don't emit ultraviolet light.

The black centre of the poppy flower stores and then distributes heat, which means that the middle of the poppy can be 2°C warmer than the outer edges of the flower. Bees and bumblebees, which have special receptors in their antennae and on their legs, are able to detect this temperature difference, which acts as an accurate guide to the pollen source at the heart of the flower, almost as if they were using a thermal camera.

Initially attracted by this excellent ultraviolet signalling system, the bees fly right up to the flower and the extra warmth then guides them to the centre, where the pollen is to be found in the darker, warmer recesses of the poppy. Pollen grains become attached to their legs or stick to their fuzzy body, and are then transferred to the pistils of other flowers, closing the cross-fertilisation loop.

SWEETCORN
Angel hair

Is sweetcorn really a flower? You'd expect to see corn cobs on a barbecue, corn flour in a tortilla or ground corn used in cattle fodder, but does corn really have flowers? A bouquet of sweetcorn flowers – really?

But why ever not?! To end up with a corn cob, you have to start with a seed, preceded by a fruit and a flower! So yes, to answer your question, before we reach the cob stage, there will have been a flower. And actually, a number of flowers, as you're about to find out.

If you were to stroll through a field of sweetcorn in spring, you'd see a number of elements on one individual sweetcorn plant: it consists of a single rigid stem with leaves at each node. The stem can grow up to 2 metres tall, which is why it's only too easy to lose yourself in a cornfield, and why they often resemble a maze. Right at the top of each stem you'll find long fine stalks emerging from the plant like a giant feather duster; these stalks carry myriad tiny flowers, but these are male flowers only. They ripen first, producing vast quantities of minuscule grains of pollen.

As you look down the main stem, you'll come across another strange flower, this time swathed in the plant's long, narrow leaves. The silks protrude from this flower, not unlike angel's hair, long plumes that grow even longer with age. These will become the future sweetcorn cob. This cob, or ear of corn, is a collection of female flowers and the long silks are pistils that are in turn connected to a great many ovaries, tightly packed inside the ear. Once they have been fertilised, these ovaries will turn into sweetcorn kernels. In other words, sweetcorn has both exclusively male and exclusively female flowers on the same plant. However, as

SCIENTIFIC NAME
Zea mays

FAMILY
Poaceae (grasses)

HABITAT
Open and sunny, plentiful water

WHERE TO SEE
These plants are among the most widely cultivated species on the planet. You can find them in fields across Europe, as well as in vegetable gardens.

FLOWERING SEASON
July

STRATAGEM

Sweetcorn flowers are wind-pollinated: they have no petals, no scent and no nectar because they don't need to attract insects. Pollen is produced by the male flowers and is collected by the female flowers with their long pistils, which are able to gather as much pollen as possible. By having flowers of different sexes, the plant maximises the chances of cross-pollination.

I'm sure you'll agree, these flowers are not particularly attractive and certainly don't have any redeeming aesthetic qualities.

You certainly won't find them in a florist's shop and pollinators are equally unimpressed. You rarely see honeybees or bumblebees hovering around sweetcorn plants. They don't give off any scent to lure in passing bees, nor do they have brightly coloured petals to divert insects off course. But this is a deliberate ploy on the part of the sweetcorn plant; like less than 10% of all flowers on Earth, sweetcorn flowers use the wind as their pollination agent – which explains why these flowers are designed as they are. First of all, to prevent the male flowers, which are located above the female flowers, from depositing pollen on flowers from the same plant, the two types of flowers reach maturity at different times: the male flowers are ready a few days earlier than the female flowers. Not all plants in the same field will start to flower at the same time, yet every grain of pollen dispersed by the wind up to 500 metres away will inevitably find silks to land on, pushing through the style and eventually reaching the ovule in the centre of the ovary, where it will deposit a sperm cell. In other words, cross-fertilisation is the norm among sweetcorn plants.

Sweetcorn flowers are designed to adapt to the wind: long stems of male flowers form a bunch known as a tassel, hanging down over the sweetcorn plants to dance more merrily to the wind's tune. Like little bells tinkling in the breeze, these flowers have drooping stamens from which copious quantities of pollen are released.

Further down the plant, the female flowers await their turn. The long silks that increase the contact surface with airborne pollen are their greatest weapon of seduction. Unlike insect-pollinated flowers, where each grain of pollen is supplied directly to the right place, i.e. on the stigma – which can therefore afford to be relatively short and protected within the flower – wind-pollinated plants have a different problem to contend with. Gusts of wind will never be as precise as insects, with the result that pollen may fall anywhere in the vicinity of the female flower. By increasing the receiving surface area in a radius around the ovaries, the female flower increases its chances of collecting pollen.

From top to bottom

Illustration from Hermann Adolph Köhler's Medicinal Guide, published in Germany in 1887.

The long stigmas, known as silks, of the female flowers prepare to receive wind-blown pollen.

The stamens of the male flowers hang down in the wind and disperse pollen.

SUNFLOWER
In full sun

Revered and eaten by the Aztecs and the Mayan people, the sunflower was only imported to Europe in the 15th century. Since then it has certainly flourished, becoming one of the world's most widely cultivated field crops. Sunflowers are grown for their edible oil, but they are also used to produce agrofuels. Yet the rising interest in cultivating field upon field of sunflowers is primarily as fodder plants: when crushed in the form of pellets or seeds, they are widely used in livestock farming. Nevertheless, sunflowers have always exerted a special kind of fascination on account of their breathtaking beauty and the impressive size of their yellow 'flowers' with their heads always facing the sun – hence the name. In fact, the flowers only track the sun when they are still in the bud stage. Once they are in full bloom, sunflowers remain in the same position, facing east, where the sun is warmest in the morning, which happens to coincide with the time at which pollinators are most active.

So we're talking about 'big yellow flowers' – or at least that's what the sunflower would have us believe! There's no doubt that, from a distance, a bee would also regard this yellow mass as a single large flower. However, on closer inspection, it becomes clear that a sunflower is in fact a collection of thousands of tiny individual flowers: between 3,000 and 5,000 of them, to be precise. Sunflowers are complex inflorescences formed by a flower head that supports all these tiny flowers making up the centre (or heart), surrounded by a ruff of different flowers with the task of supporting large petals. This is a surprisingly popular design: in fact, it's the trademark of the Asteraceae family, one of the most widespread species across the globe.

SCIENTIFIC NAME
Helianthus annuus

FAMILY
Asteraceae

HABITAT
Open position in full sun

WHERE TO SEE
Sunflowers can be found in vast yellow fields and gardens at the end of summer but are also favourites with florists.

FLOWERING SEASON
July-October

STRATAGEM

The flower is actually a collection of thousands of tiny flowers, which together form a large inflorescence. This structure enables the flower to be identified more readily by insects. The flowers face east so they can provide insects with a platform warmed by the morning sun.

It includes argyranthemums, daisies, asters and even cosmos, all of which have a similar architecture: small central flowers in the form of compact little discs (these flowers are called disc flowers or florets); they are surrounded by ray florets, each of which has a single petal.

These ray florets are the petals we pick one by one off a daisy to find out whether our love is reciprocated: they love me, they love me not... These particular flowers are sterile and form the crown of petals surrounding the middle of the flower, but if you look closely at the centre, you can see all those fertile hermaphroditic flowers that don't actually have any petals. The two sexual organs of hermaphroditic flowers are not active at the same time in order to minimise self-fertilisation. In this case, starting from the outside, you first come across the female flowers, and then the male flowers. The middle of the flower's heart will remain in bud as long as possible and the central flower will be the last to bloom. In other words, in peak flowering season, you might find seeds, fertilised flowers, female flowers, male flowers and buds all on the same flower head as you move from the outer edge of the sunflower towards the centre.

All these individual flowers produce nectar to attract pollinators, with the added bonus that the flower head is turned towards the morning sun, generating pleasant warmth for pollinators, which are cold-blooded creatures, and encouraging them to feast in the cosy conditions.

Bees are primarily attracted by the large petals of the outer flowers. They land on the edge of the flower head, then start their hunt for nectar, progressing gradually from the outer regions of the flower towards the centre. Inevitably, their journey takes them first over the female flowers and then on to the male flowers, where they become smeared with pollen. Once they reach the middle of the flower head, their job is done and they fly away, only to be attracted by the petals of another sunflower and landing, once again, on its outer rim. As it passes over the outermost flowers, the bee deposits pollen from the previous sunflower on the female flowers of the new sunflower. The result? Guaranteed cross-fertilisation.

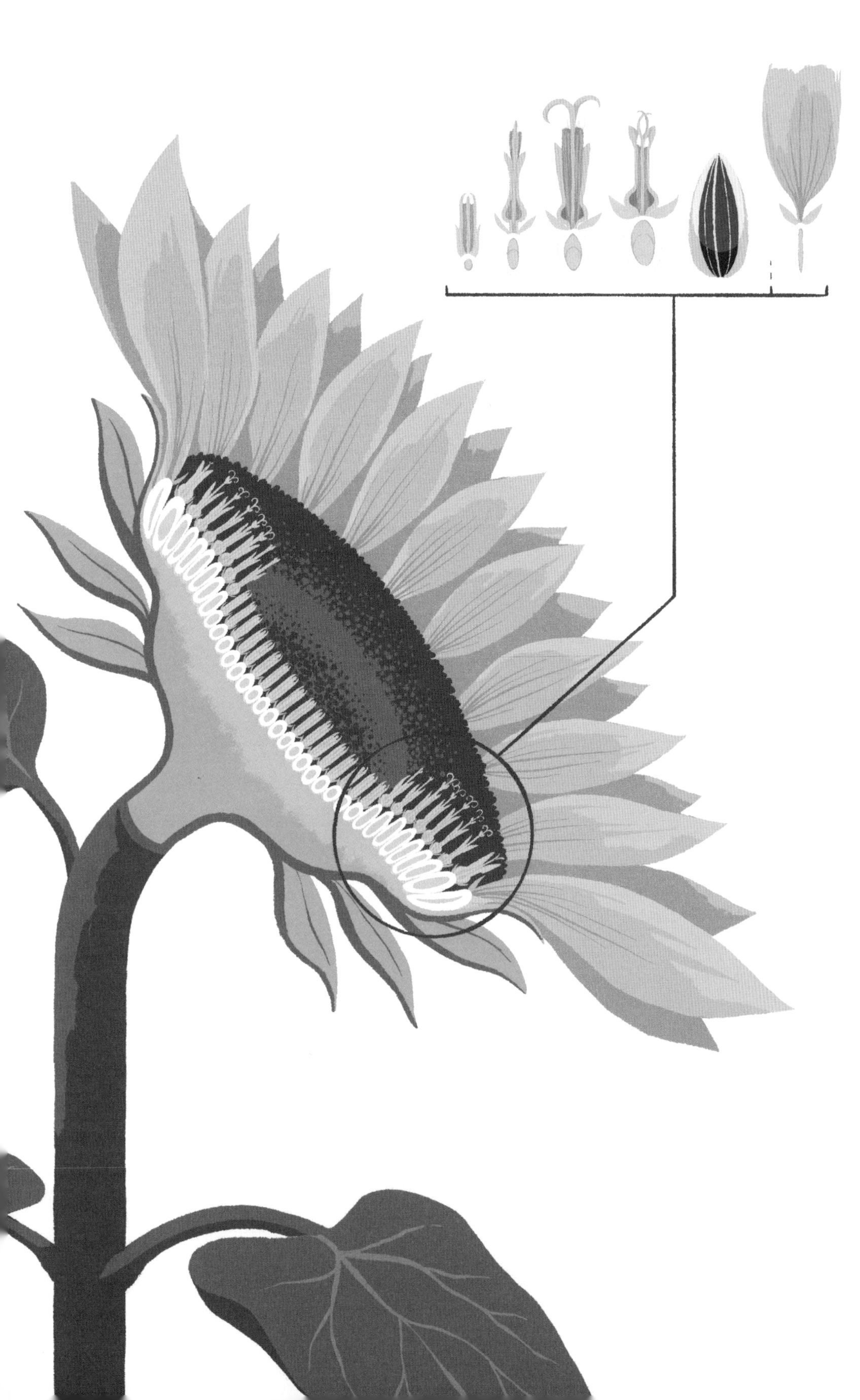

GERANIUM & PELARGONIUM

Striking natural beauty

Say the word geranium and many of us will conjure up a timeless image of window boxes perched on window sills outside quaint, half-timbered cottages with old-fashioned shutters, overflowing with cheery red flowers and green foliage all year round. In our collective imagination, 'geraniums' also go hand-in-hand with our grandmothers' well-tended window boxes and those kitsch postcards you used to see in the 1960s.

However, our collective memory leaves a lot to be desired when it comes to the accuracy of the terms used: any botanist or self-respecting gardener will tell you that these geraniums are actually pelargoniums. In botany, the word 'geranium' includes dozens of wild or cultivated species from temperate regions that are commonly found in Europe's forests and countryside. These plants have small, often pink, flowers, such as the tiny cranesbill that goes by the name of Herb Robert, found over an extremely wide area. Their flowers are made up of five petals in a circle surrounding ten stamens which are positioned around an overhanging pistil. These flowers are said to be actinomorphic, i.e. they are characterised by radial symmetry; other examples include the lily, sunflower or even the tulip.

Pelargoniums, on the other hand, are species that predominantly originate from South Africa but have become widely naturalised, selected and hybridised across Europe. Their flowers are often red or pink and they also have five petals, but, unlike geraniums, they display bilateral symmetry, i.e. they are symmetrical with respect to a central plane. These are known as zygomorphic flowers, such as snapdragons, honeysuc-

SCIENTIFIC NAME
Geranium robertianum and *Pelargonium* sp.

FAMILY
Geraniaceae

HABITAT
Herb Robert geranium: meadows, roadside verges, rocky outcrops, woodlands
Pelargonium: rocky ground, dry grasslands

WHERE TO SEE
Herb Robert geraniums (or cranesbills) can be found in the wild, growing along the roadside. There are many horticultural varieties, often used as ground cover in flower beds. Pelargoniums are also found in gardens and on window sills. They are also the world's top-selling pot plant! The French town of Bourges even plays host to a garden dedicated to the pelargonium: the National Pelargonium Conservatory.

FLOWERING SEASON
May-October

STRATAGEM

Geraniums are characterised by radial symmetry, whereas pelargoniums display bilateral symmetry. These differences aside, both flowers attract pollinators using very similar systems: a strong scent and nectar guides to tell insects where to locate the precious liquid. As soon as the insects come in search of their goal, they end up covered in pollen.

kle, lupins and the like. They have fewer than ten stamens and one pistil.

Yet despite these differences in symmetry, pelargonium and geranium species are very similar in evolutionary terms, which is why they are members of the same family, that of the Geraniaceae.

Both flowers use the same stratagems to alert pollinating insects of their presence because they depend on these creatures to ensure cross-fertilisation. First of all, they use fragrance so that insects can identify them from afar. After that, their beautiful contrasting flowers catch the insect's eye. As the insects get closer to the plant, more subtle signals take over and draw them in: these are the nectar guides. Whether these are the lighter dots in the centre of the circular corolla found in the Herb Robert geranium or the distinctive markings on the upper petal of some pelargoniums, in both cases the flower leaves the insect in no doubt where the nectar can be found: right in the middle.

The Geraniaceae family have a system whereby the male and female parts of the flower mature at different times. All the flowers start off male, then the female organ takes over as it unfurls and becomes receptive. This ensures that the pollen from one flower cannot come into contact with the pistil of the same flower, but pollination will take place if a pollinating insect has visited a young flower before feeding from an older flower. In this case, it will probably have accidentally picked up a small amount of pollen on its furry body and these tiny grains will then be deposited on the receptive stigma of the older flower. This is how cross-fertilisation works with geraniums and pelargoniums.

Pelargonium flowers are exceptionally long-lived. Unfortunately, it seems that the horticultural selection process that has led to the flowers becoming virtually everlasting has also deprived the plant of a flourishing relationship with potential visiting insects. A study assessed the attractiveness of 30+ traditional garden flowers from various cultivars selected for their horticultural uses and their ability to please the human eye. The aim was to establish whether bees have the same tastes as humans and whether our view of nature is the same as the bee's-eye view. The results of this study indicated that the pelargonium is one of the flowers that insects visit most rarely, to the extent that it might not even welcome any visitors at all for days on end, whereas the majority of other flowers were visited hundreds of times a day by a range of different insects. By bending Mother Nature too much to our own tastes, we are at risk of scaring away Nature's main benefactors.

From top to bottom

A pot of geraniums, Odilon Redon, 1905.

An ant feeding on nectar from Geranium Rozanne (*Geranium himalayense x wallichianum*). The flower displays radial symmetry.

Pelargoniums have a long flowering period but attract very few insects.

COWSLIP

The long and short of it

A common sight in springtime, little yellow flowers nodding in the wind, for all the world like jaunty bells perched atop narrow, light green stems covered in downy hairs. This is *Primula officinalis*, commonly known as the cowslip. The primula or primrose is widespread throughout temperate regions and can readily be found in meadows, on forest margins or in fields. They are among the earliest plants to flower in spring, as suggested by their alternative Latin name *Primula* (first of all) and *veris* (relating to spring).

Primroses have been remarkably successful in tackling the problem of self-fertilisation; this leads to reduced genetic diversity within the population and ultimately causes an ever-growing number of mutations with a detrimental impact on the species. As with most flowers, cowslips need pollen from another flower to be deposited on the pistil, no matter whether the other flower is just ten centimetres or thousands of metres away. The most important thing is that the pollen must not come from the same plant, because this would mean that the flower is fertilised by a sperm cell with the same gene pool as the ovule. Especially as cowslip flowers are hermaphroditic: they have both male and female reproductive organs.

These little yellow flowers are full of nectar, making them irresistible to insects, especially honeybees and bumblebees, but also solitary bees such as mason bees, which develop into full-grown adults in the early days of spring.

Initially, the subtle fragrance of the cowslip stimulates the bees' antennae, followed by the little patches of yellow clustered at the top of the flower stalks,

SCIENTIFIC NAME
Primula veris

FAMILY
Primulaceae

HABITAT
Roadside verges, banks, hedgerows

WHERE TO SEE
You can find cowslips on roadside verges in limestone areas, in meadows, pastureland and sunny woodland glades. Primroses are domesticated cowslips, commonly found in gardens or on balconies.

FLOWERING SEASON
April-June

STRATAGEM

The plant has two types of flowers: some have long stamens and a short pistil, whereas others have short stamens and a long pistil. If a bee visits a flower with long stamens, it can only pollinate a flower with a long pistil, and vice versa. This clever design trick reduces the risk of self-fertilisation.

above the surrounding grass, but sufficiently bright to catch the insects' attention.

The bees plunge deep into the flowers to feast on their pollen and nectar booty, then move on to other flowers, without any idea of their genetic make-up.

It's entirely possible that, in inserting its head inside the tube-shaped corolla of the cowslip, a mason bee might deposit pollen on the pistil of the same flower, leading to self-fertilisation, the very thing the plant is trying to avoid. Disaster!

But thanks to evolution, the cowslip has developed an ingenious ruse to prevent this happening. Take a look inside the tubular form of a number of different cowslip flowers and you'll see that these flowers have two different set-ups. On the one hand, some have a long pistil, extending beyond the corolla, but with very short stamens – the pollen is only to be found at the base of the flower, just before the flower narrows, leading to the nectar reserves right at the bottom. This type is known as a longistylous flower (with a long style). The other half of the cowslip population has flowers with the opposite design – these are known as brevistylous (short-styled) flowers. They have a short pistil, which just protrudes from the constriction before the nectar reserve, and very long stamens which almost extend beyond the tube formed by the fused petals.

If a mason bee (or any bee, for that matter) decides to come and feast on the nectar at the base of a cowslip flower, it will have to push its head right down to the base of the flower, stick out its tongue and insert it through the narrow part in order to reach and sip the nectar at the bottom. If the flower is longistylous, the pollen will be right at the base of the flower and the insect will emerge with its head, and only its head, daubed with pollen. After visiting this first flower, the mason bee continues its journey to another flower. If, this time, it finds a brevistylous flower and pushes its head into the depths of this flower, the pollen attached to the bee's head will come into direct contact with the pistil, which protrudes just above the nectar reserve in this particular flower structure. The flower can therefore be fertilised, which is exactly what the plant wants, because a single cowslip plant will never bear both brevistylous and longistylous flower types. In other words, cross-fertilisation can take place. Once the insect is inside a brevistylous flower on the hunt for nectar, it will wriggle around, moving its legs and body in the process, causing these to become covered with pollen, which is in this case located very high up inside the flower. This pollen will only come into contact with a pistil that is also in an elevated position, in other words in a longistylous flower. Game, set, match, cowslip! Cross-fertilisation has become the norm.

BROADLEAF PLANTAIN

Blowin' in the wind...

Although widely found in the countryside and with a tendency to encroach on our lawns, the plantain is clever enough to conceal its presence until the dog days of summer, when its long stems bearing unusual-looking flowers allow it to throw off its cloak of invisibility. Yet this plant is so humble and so fragile that it meekly grows where other plants would rather avoid – in places where it is liable to be trampled, either by cattle or by humans walking on pathways. The name plantain is said to originate from its habit of growing in areas frequented by foot traffic, or from the shape of its leaves, which grow close to the ground and are said to resemble the sole of the foot. However, there's no doubt that trampling, although it does destroy a good many plants, works in the plantain's favour, giving it more space to grow, and draw water and essential growth nutrients from the soil without fear of competition.

The plantain is widely distributed and has leaves that are often rounded, depending on the individual species, but always with parallel veins. When the time comes, the plantain uses all its strength and energy to send up a tall flower spike, as high as it can grow, comprising many flowers, from the middle of its leaves, which are arranged in a rosette growing at ground level. The flower spike is an inflorescence made up of a great many tiny, tightly packed flowers. These individual flowers are not usually particularly attractive: in fact, they're quite plain in both shape and colour, not unlike ears of wheat.

But if you look closely at the plantain flower spike, it soon becomes evident that it has no garish petals, no scent and not a sign of nectar. What kind of insect or bird would be tempted by this strange kind of flower?

SCIENTIFIC NAME
Plantago lanceolata

FAMILY
Plantaginaceae

HABITAT
Meadows, fields, lawns

WHERE TO SEE
Found in cracks in paving in urban settings, or on lawns or country lanes.

FLOWERING SEASON
March-October

STRATAGEM

The flowers are wind-pollinated. The plantain carries its inflorescences as high as it can, with flowers that are successively female then male, to ensure that the pollen is dispersed as far away as possible. The male flowers are positioned below the female flowers to prevent pollen from the same plant falling on its own female flowers.

The plantain is not the sort of plant to seek out the company of other species; oh no, plantains are very much wind flowers. They strive to throw up their flower spikes as tall as possible on long, bare stems dancing in the wind for the sole purpose of putting as much distance as possible between them and the ground to make sure their pollen is dispersed far and wide.

Let's look in more detail at a flowering spike to find out how pollen is produced. Select any flower spike at random and it won't be unusual to find flowers in various stages of maturity. In fact, the flowers mature in a very specific order: they flower from the bottom up, which means that the buds at the top of a flower spike will be the last to open. During the initial flowering stage, only the female part of the flowers will be active: the pistil will be receptive to any pollen brought on a gust of wind. The flowers then become male, which means the pistil is no longer receptive; finally, once the stamens have dispersed their pollen, they wither and fall, after which the fruit and the seed are able to develop.

Looking from top to bottom, we can see a first section comprising flower buds, then a section with female flowers and their white pistil, looking for all the world like a little feather duster on the flowering spike. This is followed by a section with male flowers bearing long stamens with very fine anthers culminating in filaments that vibrate in the wind. The slightest breath of wind will pick up the pollen and disperse it in the surrounding area. The next stage consists of fertilised flowers that are starting to form seeds.

The plantain also uses the wind to disperse its seed, scattering it over lawns and roadside verges. Despite being a very common plant, the plantain is no stranger to culinary or medicinal uses: you can eat the flower spikes, tossed in butter or pickled, while the leaves have antihistamine properties and are ideal for treating the unpleasant itching that often accompanies insect bites. Next time you're bitten by a mosquito or stung by a bee, try crushing a young plantain leaf between your fingers and pressing it against the affected area. Any itching and stinging sensations should soon disappear.

From top to bottom

Botanical plate, 19th century.

The plantain's leaves are said to be lanceolate, or lance-shaped.

The male flowers on the flower spike can be identified by their dangling stamens.

VIOLET
Anything for a quiet life

Violets make their appearance in springtime, discreet yet omnipresent on roadside verges, in fields or deep in the woods. They have a sweet, yet subtle fragrance, although it often takes several attempts to appreciate it properly. This is because the scent of violets has an unfortunate tendency to saturate our olfactory receptors, which only regain their function after a slight delay.

Violets are small flowers made up of five axially symmetrical petals that act as flags for pollinating insects, like bees, guiding them towards the centre of the flower, which is usually a paler colour. This is where the flower keeps its nectar reserves in a spur pointing towards the rear of the flower. As insects insert their tongue into this spur, they end up with grains of pollen on their heads, which they go on to deposit on another flower.

And yet violet flowers are quite discreet, sometimes a little too discreet, especially in springtime when there are many other flower species clamouring to capture the attention of all kinds of passing insects. So there is a significant risk that violets might be overlooked in the pollination party, jeopardising the long-term survival of the species. Let's not be too hasty though: these tiny flowers aren't quite that short-sighted. Even though cross-pollination is their preferred mechanism, violets don't expect insects to do all the work. On the principle that if you want something doing, do it yourself, violets produce strange sorts of hidden flowers towards the end of spring. They have all the usual components: sepals, petals, stamens and a pistil. The only difference is that nothing is visible. The flower remains in

SCIENTIFIC NAME
Viola odorata

FAMILY
Violaceae

HABITAT
Meadows, roadside verges, hedgerows, gardens

WHERE TO SEE
Violets can be grown in gardens, alongside their cousin, the pansy. They can be eaten in crystallised form or as sweets known as Parma violets. The French town of Toulouse is famous for both the sweets and the national collection of violets grown in its municipal glasshouses (Conservatoire National de la Violette). Liège in Belgium is also renowned for its violet-flavoured boiled sweets.

FLOWERING SEASON
February-May, but sometimes also August-October

STRATAGEM

Some flowers are fertilised by pollinators like bees, but at the end of the season some flowers never seem to open, remaining in the bud stage even though their pollen and pistils are mature. In this case, fertilisation takes place inside the bud.

bud form, with the green sepals providing a protective shell around the precious sexual organs. These pseudoflowers continue to maturity and once the resulting pollen grains are ripe, they can easily be deposited on the adjacent pistil.

This all happens secretly, hidden away from greedy insects, in a process known as cleistogamy (from the Greek *kleistos,* meaning closed, and *gamos,* meaning marriage). The benefit of this system is that there's no need to produce that valuable commodity, nectar, nor to manufacture large amounts of pollen. The downside is that, in the long-term, self-fertilisation does not allow genetic mixing and is likely to perpetuate detrimental mutations within the population.

In other words, in the violet family, it all comes down to compromise. If, despite the many tempting offers from other flowers, the violet manages to attract pollinators, cross-fertilisation may take place, but, to be on the safe side, seed production is always guaranteed, whatever happens, thanks to this magic trick involving pseudoflowers.

Seed-based propagation is not the violet's primary strategy in any event: like the strawberry plant, violets are also able to use a propagation technique based on cloning (also known as vegetative propagation or asexual reproduction) by means of stolons. These are runners that emerge from the base of a violet and take root a few centimetres away from the main plant.

Their ability to reproduce readily explains why some violet species, such as the sweet violet, or *Viola odorata*, are so easy to grow and distribute commercially. Their unique fragrance has been enjoyed by many different civilisations over the years, including the Romans, who appreciated the violet for its aphrodisiac and medicinal properties. Then there were the residents of Toulouse, who were driven by the violet's culinary potential: during the 19th century, many market gardeners in the Toulouse area cultivated violets for posies almost all year round, even sending them as far afield as Russia. The flowers were also used to make cordial and liqueurs, but the main use of the fresh flowers, harvested in March, entailed crystallising them in sugar. Nowadays, Toulouse is renowned across the world for its confectionery: the famous Parma violets.

Other violet species are also grown in the United States as flavourings for candy, not least marshmallows, which feature on the list of confectionery created by Willie Wonka and his chocolate factory in Roald Dahl's well-known book.

MYOSOTIS
Forget-me-not

The little blue and yellow flowers of the myosotis, commonly known as the forget-me-not, are modest and unassuming. They don't seem to mind where they grow, be it on riverbanks, in woods or meadows where they create a shimmering blue haze through the flowerbeds. First-class strategists that they are, forget-me-nots are the kings of visual communication, banking on their contrasting colours to attract or repel potential pollinators. Bees, bumblebees and flies or other members of the Diptera order are frequent visitors, stopping by to feast on nectar or gather a small amount of pollen from the five sacs attached to the base of the five fused petals making up the corolla. There is a tiny hole in the middle of the ring of pollen, through which pollinators can access the nectar, at the same time depositing pollen, which will come into contact with the pistil.

However, this strategy doesn't always work: nectar is only produced when the flower is mature but hasn't yet been fertilised. It isn't in the forget-me-not's interests to produce more nectar than absolutely necessary if the pistil or the stamens aren't yet mature. And once fertilisation has taken place, the plant no longer needs to attract bees. From the pollinators' perspective, there's not much point going to check out a flower that's too young or one that's already been pollinated.

This is why the forget-me-not, that masterful Impressionist painter of the garden world, has developed a kind of code: an effective visual system that makes use of its predominant colours. If a flower is mature and producing nectar, it will be blue with yellow pollen at its heart. However, juvenile flowers, which are not yet producing nectar, will have pink petals, but the

SCIENTIFIC NAME
Myosotis sp.

FAMILY
Boraginaceae

HABITAT
Banks, woodlands, roadside verges, meadows

WHERE TO SEE
Many ornamental varieties are cultivated in gardens.

FLOWERING SEASON
April-June

STRATAGEM

The flowers are blue with a ring of yellow pollen when nectar is available. If the flowers aren't mature enough and there isn't any nectar yet, the petals will be pink. Once the nectar supply is exhausted, the petals will be blue, but the ring of pollen will turn white. This clever system tells the pollinators whether or not they can expect to find any dinner.

pollen will already be yellow.

The flower's ability to change from pink to blue is thought to be due to the pH of the petals: juvenile flowers tend to have acidic petals (i.e. pink), whereas mature petals are more alkaline and therefore blue. When the flower is starting to fade, no longer has any pollen left, and its nectar supplies have dried up because it has already been fertilised, the petals will remain blue, but the ring of pollen in the centre will become paler, verging on white.

Pollinators merely have to use their eyes and remember the simple yet effective colour code. You only have to look at a bed of forget-me-nots to realise that this colour-coded system really does work: insects are much more likely to visit blue flowers with a bright yellow centre. This is because bees and bumblebees are quick learners and soon understand that a particular shape, colour or scent is associated with a reward. They retain this knowledge over a long period until all the forget-me-not flowers have turned white in the centre. Many pollinators have the cognitive ability to remain loyal to a particular flower species throughout its flowering season. This is known as floral constancy.

Nevertheless, you might be wondering why forget-me-nots keep their flowers in situ for so long once they've been fertilised: it all comes down to publicity. Making yourself visible at all costs, from as far away as possible! Truth be told, in the race to attract insects to your species, the more you put yourself out there, the more successful you will be. A mass of flowers will appear more attractive to the compound eyes of a Hymenoptera than one or two individual flowers. Members of the Hymenoptera order have very good colour vision and can detect blue and pink particularly accurately, enabling them to distinguish effectively between these two colours even though they can appear very similar to the human eye. However, when they are more than 50 cm away from a group of flowers, they are unable to tell the difference. The forget-me-not therefore uses two visual signals from its bag of tricks: first, it needs to attract insects from afar, so all its flowers have a role to play. In any event, bees and bumblebees are unable to distinguish fertile flowers from infertile flowers at this distance so they are likely to come anyway. Then, from close up, the flowers with a white centre are no longer needed because insects will only visit blue flowers with a yellow ring around the middle. But by that stage, the forget-me-not's trick has served its purpose.

From top to bottom

The inflorescence of this forget-me-not is made up of mature flowers as well as a few flowers that no longer contain any nectar or pollen: the corolla, which is usually yellow, has turned white.

Macro shot of honeybee feasting on nectar from a cluster of forget-me-nots.

Forget me not (Myosotis alpestris), Mary Vaux Walcott, 1907.

NETTLE
A sting in the tail

They drive you mad, they sting, they itch – they're a pain, in other words. 'Oh, but they're good for the circulation,' you might hear people say. Thanks, but no thanks... Nettles are stinging plants with the scientific name *Urtica* from the Latin word for sting. The sting is the plant's defence mechanism: its leaves are covered with tiny hairs, which, when they come into contact with another surface, like human skin, disperse into jagged fragments of silica that cut the skin as effectively as glass. The base of each stinging hair also releases a cocktail of unpleasant substances including formic acid and histamine. This is how the plant protects itself from herbivores – and the more it is attacked, the more it reinforces its defence mechanisms. Once a plant has been nibbled by a herbivore or trampled underfoot, the quantity of irritants released increases next time round.

No wonder *Urtica* triggers such strong reactions. From an aesthetic perspective, the nettle isn't much to look at; however, the way it is pollinated is extremely interesting. Nettles don't have showy flowers, nor do they release any scent as such, but they do have small clusters of tiny flowers of the kind you might see on some trees, like hazels. And, just like hazels, nettles have small, scentless, green flowers arranged in catkins nodding in the wind: yes, you've guessed it, that's because they are in fact pollinated by the wind. These plants are anemophilous, in other words.

If you look carefully, you can see that some plants have the same kinds of catkins, while others have different ones. Look closer still – making sure you don't get stung – and you'll notice that some of these tiny flowers produce pollen, whereas others have a pistil.

SCIENTIFIC NAME
Urtica dioica

FAMILY
Urticaceae

HABITAT
Plains, wasteland, urban environments, gardens

WHERE TO SEE
Extremely widespread; nettles grow almost everywhere.

FLOWERING SEASON
June-September

STRATAGEM

The nettle is exclusively wind-pollinated. Some plants have only male flowers, while others have only female flowers to prevent pollen from the same plant landing on their own pistils. Once the male flowers have dried out, they swiftly dispatch grains of pollen up to several metres from the plants to encourage wind dispersal.

If you didn't know, you'd never be able to guess: the fact is that some plants have male flowers, while others have female ones. These non-hermaphroditic species are also referred to as dioecious species, as the nettle's botanical name might suggest: *Urtica dioica*.

One thing's for sure: a flower is either one sex (stamens in the case of male flowers) or the other (pistil in the case of female flowers). And there will only be one type of flower on the same cluster (or catkin).

In very rare cases, nettles may have both male and female flowers on the same plant, but the flowers are still not hermaphroditic. In this case, the plant is said to be monoecious.

Male plants do not necessarily grow alongside female plants; indeed, very often, male flowers may be quite a distance from their female counterparts. A good gust of wind should blow pollen between these clumps of nettles, but if the wind isn't strong enough on any given day, the nettle has another trick up its sleeve to send pollen as far away from the male flowers as possible: it has a foolproof strategy in the form of a pollen catapult. As they start to flower, the stamens of the male flowers are curled up, protecting their precious cargo of pollen. The anthers dry out as they ripen and, when they are fully mature, the cells of the anther walls shrink under the sun's rays, creating tension. It takes them less than a second to eject all their pollen grains in a cloud of dust, which is soon whirled up and carried away by the wind. Male flowers produce huge quantities of pollen and some of these grains will undoubtedly end up on corresponding female flowers, leading to cross-reproduction, even though some pollen may find its way into human nostrils, contributing to that oh-so-common complaint of hay fever.

BEE ORCHID
Sleeping beauty

Imagine a lovely, long weekend in May. You've just had a delicious lunch and now you're out walking in the charming countryside of the Périgord Vert region in the northern part of the Dordogne – or any other limestone plateau for that matter. There's nothing quite like a breath of fresh air... If you're lucky, and keep your eyes peeled, you might well come across this treasure as you scan the verges beside the path.

At first glance, you might think you've spotted a bee perched on top of a blade of grass, pondering on life's mysteries and the fate of its fellow bees under increasing pressure from pesticides. But as you creep closer, keen to inspect this creature, you realise that the supposed bee, or bumblebee, or whatever it is that you were expecting to see flying off, buzzing angrily, isn't moving at all... Very odd. It's not moving because it isn't an insect at all. It's a flower. To be precise, it's a bee orchid, one of the many orchid species found in the French countryside.

And if you've fallen for its age-old game, you can rest assured you won't be the only one! Imagine how a male bee, training its gaze full circle over the surrounding vegetation, could equally well be taken in and draw closer to investigate this insect with its tiny wasp waist and swollen abdomen, dozing away so peacefully in this bucolic idyll – entirely understandable, in fact.

And that's the strategy in a nutshell: the bee orchid has a very unusually-shaped lower petal, otherwise known as a labellum or lip: bulging and well-rounded, it resembles the abdomen of a bee. It even has the same colouring.

If that doesn't raise an eyebrow, nothing will.

SCIENTIFIC NAME
Ophrys apifera

FAMILY
Orchidaceae

HABITAT
Meadows, roadside verges, hedgerows

WHERE TO SEE
The bee orchid is quite rare and can be seen during country walks, on the edges of woods or in upland regions like the Massif Central.

FLOWERING SEASON
April-July

STRATAGEM

The flower resembles the black, hairy body of a bee and gives off scents similar to those of a female of the species. It attracts male bees which are keen to copulate with their female counterparts. As part of this pseudocopulation process, the males become coated in pollen, which they then transfer to the next flower they visit.

Interestingly enough, the *Ophrys apifera* (*ophrys* means 'eyebrow' in Greek, because as well as being swollen, the labellum is hairy, enabling it to mimic the insect's body even more successfully; and *apifera*, from the Latin term *apis*, or bee, and *fero*, carrying) uses a reproductive strategy based on the male bee's desire to find a mate.

The bee orchid does everything in its power to mislead male bees, which are understandably rather distracted. Its deception takes the form of promising colours and shapes, but also extends to the use of scents and even pheromones. These are chemical stimuli that imitate the sexual pheromones of a female bee and are released into the air by the orchid to titillate the antennae of male bees in the vicinity.

Dressed up to the nines and smelling divine, the flower looks for all the world like a female bee, fast asleep on a flower stalk – to the eyes and antennae of a male bee, at least.

Only solitary wild bees are taken in by this ruse, however. One species most likely to rise to the bait is the *Eucera longicornis*, or long-horned bee. The male long-horned bee tries to mate with what it thinks is a female of the species. During this act, described rather prosaically as pseudocopulation, it ends up bashing its head on the upper petals, where two sacs of pollen, known as the pollinia, are strategically located. On coming into contact with the insect, the grains of pollen are transferred from the flower to the visitor's furry head. They are smeared with a kind of glue, which sets fast as soon as it touches the bee, securing them like a second pair of antennae on the bounty hunter's forehead. Job done, the male bee withdraws, flies on for a while, then is once again tantalised by another enticing scent; swerving sharply, it finds another potential lover in a heartbeat. In its ardour, it hastens towards its new ladylove, the pollen glued to its fur flattened by the speed of its approach. During the act itself, the bee again nudges its head against the top of the new flower, where the pollinia are located, but also home to the pistil.

From top to bottom

The rare bee orchid can be seen in meadows, on roadside verges and in hedgerows.

Botanical plate, 20th century.

The purple-hued lip takes on the appearance of a female long-horned bee, even down to its hairy body.

This time the pollen is detached from the hairs by mechanical pressure and deposited on the pistil. And so the story comes to an end: the grain of pollen moves down the pistil, attracted by chemical signals given off by the ovules, releases its sperm cells, and before you know it, the plant has been fertilised.

Unbeknownst to our poor bee, the bee orchid's pollination strategy has paid off.

BURDOCK
Natural Velcro

Burdock is a very common plant on the continents of Europe and Asia. It can grow up to two metres tall with large rhubarb-like leaves. Both its stems and its roots are classic ingredients in Chinese cuisine, in particular. They produce very characteristic flowers, although they also bear a passing resemblance to other well-known species, like thistles.

Right at the top of these statuesque plants in their typical meadow or streamside habitats, you'll find little reddish-purple tufts, almost like miniature shaving brushes, emerging from a prickly globe-shaped bud. This arrangement is not dissimilar to other plants like thistles or cirsiums, although these have much less accommodating leaves than the large, soft, hairy leaves of burdock plants.

Yet the flowers of thistles, and even knapweed (*Centaurea*) certainly have a similar basic structure to the burdock. The 'flower' you can see, and which insects most certainly recognise as a single flower, is actually an inflorescence, or mass of tiny, densely-packed flowers. This is what's known as a capitulum, or flower head, and it has a similar arrangement to the sunflower, or dandelions and daisies for that matter. And yes, you've guessed it, all of these flowers belong to the same family: Asteraceae.

Unlike the sunflower, however, the burdock doesn't have outer sterile flowers bearing petals, but a bunch of small flowers primarily consisting of a long tube formed of six very narrow, fused petals, which, in the bud stage, resemble long hairs, not unlike the hairs of that shaving brush we mentioned earlier.

While the capitulum is still in bud, an ingenious mechanism is preparing for action inside the flower;

SCIENTIFIC NAME
Arctium sp.

FAMILY
Asteraceae

HABITAT
Wasteland, meadows, hedgerows

WHERE TO SEE
Burdock is a common plant that grows in small clumps or is found alongside hedgerows. In winter, its prickly burrs make it easy to identify.

FLOWERING SEASON
June-August

STRATAGEM

The flower is made up of thousands of tightly-packed flowers in a flower head forming a small tuft. When the flower first opens, pollen is collected from the base of the pistil to expose a larger surface area and thus maximise the amount of pollen deposited on insects. As they develop, the flowers are alternately male and female to prevent self-fertilisation.

the pollen carried on the stamens, forming a sleeve around the style, is already ripe. The style growing inside the mass of pollen will become coated with the powdery grains. When the flower opens, it looks like a long, off-white style with a stigma in three separate sections and, at its base, a tuft of pollen-smothered hair in a fetching shade of pink. Underneath, you can still see a dark purple sleeve of pollen. This method, known as secondary pollen presentation, is fairly common among the Asteraceae family – such as sunflowers – but it can also be observed in campanulas (bellflowers).

To avoid cross-pollination, the flowers start off male. Then, a few days later, the stigma becomes receptive and the flower becomes female. Deep within the flower, nectaries produce nectar to the delight of visiting bumblebees, butterflies and a variety of other insects, which dart from flower to flower, only to become inadvertently daubed with pollen, and deposit this valuable resource on older flowers, now receptive to this pollen, perpetuating cross-fertilisation in the process.

One little fruit fly, *Tephritis bardanae*, doesn't just stop there: when it arrives at the host flower, it takes a sip or two of nectar, dispersing the pollen at the same time, but the biggest attraction for this fly is the flower head itself, where the female lays its eggs. The developing larvae can then feed on the plant's developing seeds.

Below its fluffy floral top knot, the burdock flower is surrounded by a fortress of prickles: hundreds of fused bracts (which are actually a form of leaf, usually found at the base of flowers) form the spiky lower half of the flower. Each bract ends in a long hook: when winter comes and the flower has long since faded, but the seeds are ripe for dispersal, the hooks of these bracts are ideal for hitching a lift on the fur of any large mammals unfortunate enough to pass their way – dogs are frequent victims, but also horses with their long, swishy tails, to say nothing of children's hair… This impressive burr was the inspiration behind the invention of Velcro.

Next page (double spread)
Burdock plant growing alongside a field of wheat.

LOVE-IN-A-MIST

If you want something doing, do it yourself...

Love-in-a-mist (or *Nigella damascena* to give it its official name) is a familiar garden plant, known for its blue, white or pink flowers with a star-like ring of petals around its prominent crown of stamens, encircling four ovaries with long styles protruding from the very heart of the flower. The nigella plants cultivated in the Middle East and in Asia (*Nigella sativa*) are renowned for their seeds, which are used as a spice (also known as black cumin). Farmers will be familiar with another species, the field nigella (*Nigella arvensis*).

These flowers rely primarily on bees, bumblebees and other pollinating insects to come calling and provide their pollination services, but, if push comes to shove and no-one turns up, they've worked out ways of managing without...

The flower is discreet, but stands out from its surroundings by its shape, the vivid blue of its petals and its many stamens. It issues an open invitation to insects keen to fill their boots with nectar and take away a bonus dusting of pollen at the same time, as is often the case where flowers rely on insects to assist with pollination. The secret is knowing where to find the nectar: it's often hidden deep inside the flower, protected from predators that might not be as willing to play their part in the pollination game. So how do insects find the nectar if it's hidden from view? And what looks more like nectar than nectar itself? Scientists have discovered that love-in-a-mist provides fake drops of nectar at the bottom of the flower to guide bees towards its treasure trove and signpost the entrance to the hidden nectar. Between each petal there are two glistening protuberances that emit ultraviolet light – which Hymenoptera can easily

SCIENTIFIC NAME
Nigella damascena

FAMILY
Ranunculaceae

HABITAT
Wasteland, rocky outcrops, agricultural land

WHERE TO SEE
Although it can be seen growing in gardens or on the edges of fields, love-in-a-mist is cultivated as a crop in the Middle East.

FLOWERING SEASON
June-August

STRATAGEM

If the flower hasn't been visited by a pollinator by the end of its flowering life, its pistils carry on growing until they reach the stamens and seek out pollen so they can fertilise themselves. In other words, come what may, the love-in-a-mist will form seeds, even if those resulting from self-fertilisation aren't quite as high quality.

detect – looking exactly like two droplets of nectar; these protuberances are known as pseudonectaries.

So, in the first few days of flowering, the pollen is ripe and ready to be deposited on insects' backs, but the pistils are not yet mature. A bee taking its fill of nectar will soon find its head or back liberally dusted with pollen. It will gather even more nectar as it continues its journey before eventually returning to the colony. As it lands on another love-in-a-mist flower, it will use the pseudonectaries to locate the genuine nectar and, as it does so, if the next nigella flower is a little older, it will deposit pollen on the stigma. This is cross-pollination in action.

But sometimes bees are few and far between, for instance when summer is coming to an end and there are no insects around to deposit third-party pollen on the stigmas of a love-in-a-mist flower. All is not lost, however. Although it's certainly not the ideal solution, our love-in-a-mist could still save the day and switch to self-fertilisation if all else failed. As time goes by, the many pistils at the centre of each flower carry on growing until they eventually start to bend and before you know it, they have almost reached the stamens. With a little more effort – a matter of days, no more – they have achieved their goal: the pistils have grown throughout the summer, so the stigmas are now physically touching the pollen-bearing stamens of the same flower.

Self-fertilisation is the inevitable result, meaning that any nigellas that haven't managed to attract bees will still go on to produce seeds. Remember, though, that as a general rule self-fertilisation is much less viable in the long term and a self-fertilised love-in-a-mist flower will not only produce far fewer seeds, but any seeds it does produce will also be much more likely to suffer from degeneration. That said, as an outcome, it's got to be better than not reproducing at all...

From top to bottom

Portrait of a love-in-a-mist flower, Emile Gallé (studio), undated.

The delicate love-in-a-mist has many stamens and green pistils, which blend in with its leaves.

The pistils of this nigella grow until they reach the pollen.

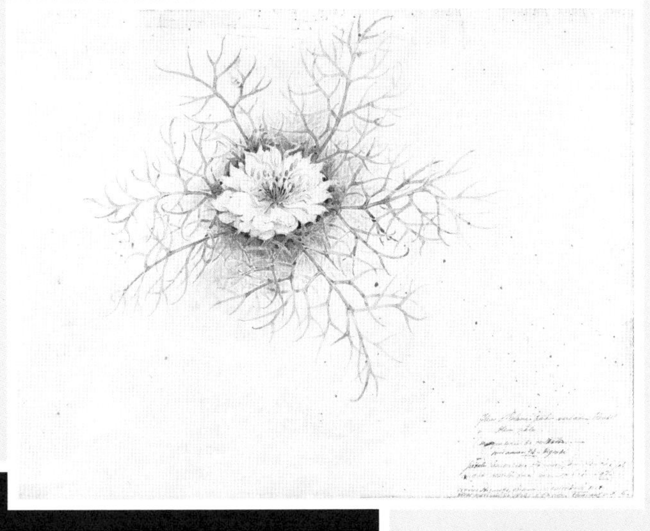

WILD CARROT
An open bar

Everyone's familiar with carrots, those bright orange root vegetables renowned for helping you see in the dark – if legend is to be believed. The French even maintain that eating carrots keeps you sweet! Just like any cultivated plant, the carrots we see in the supermarket are the result of selecting wild plants, then crossing them with each other over thousands of years – for as long as humans have been on Earth. As promising seeds were identified, they were kept and re-sown from year to year, eventually leading to a plant with the desired properties: large, fleshy roots in this particular case. This is known as domestication.

The wild flowers of the domestic carrot are still to be found in fields, meadows and roadside verges. Their large inflorescences are easy to identify, with their off-white parasols held above tufts of feathery leaves. Look closer, though, and you'll be able to appreciate their subtle architecture: tiny, delicate white flowers arranged in circles of varying sizes, forming flattish heads, not unlike lace doilies. These collections of miniature flowers are known as umbels and they are the chief characteristic of a large family of flowers that goes by the name of umbellifers (or Apiaceae). This family includes fennel, hogweed and angelica.

The umbels of the wild carrot (also known as Queen Anne's lace) consist of small white flowers, often with darker purple or black flowers right in the middle. These flat umbels are open to the wind and easily accessible by any passing insect, whether arriving by air or crawling up the stem. From afar, the mass of individual flowers is easy to spot and the inflorescence serves as an immediate marker indicating the presence of a food source.

SCIENTIFIC NAME
Daucus carota

FAMILY
Apiaceae

HABITAT
Meadows, banks, wasteland

WHERE TO SEE
The wild carrot is a very common plant and can easily be spotted on walks in the countryside as well as in more urban settings. Carrots are cultivated plants, well known to gardeners; the French town of Angers even holds the national carrot collection.

FLOWERING SEASON
May-October

STRATAGEM

Carrot flowers are tiny but clustered together in large plate-like structures, known as umbels, which act as insect landing pads. Right in the middle is a black flower not unlike a feeding insect, a strategy that seems to attract other insects. The carrot takes a very undiscriminating approach to pollination as its flat umbels attract many different species of insects.

Their main attribute is that they make excellent landing pads: despite their lacy appearance, umbellifers aren't particularly fussy and are happy to act as public heliports, welcoming any relatively skilled aerial acrobat that chances to crash-land with a modicum of grace on their flower plates. And to encourage the notion that their flowers are a fruitful animal feeding ground, it seems that the dark blooms in the centre of each flower head are designed to attract insects: from afar, those dark central flowers look like an insect in mid-banquet – at least if you're a short-sighted beetle! Pollinators are often swayed by mimicry and if they see another insect on a flower, they can't resist following suit on the basis that the food must be good if another customer is already tucking in...

Once on the flower, insects of all kinds – and by this we mean many different species, ranging from beetles that munch on the flowers, slurping up the nectar as long as it's not too far down, to bumblebees, more skilled in the acrobatic arts, not to mention flies, gnats and butterflies – feast to their heart's content on nectar and pollen. The umbels contain either hermaphroditic flowers (which have pistils and stamens) – these are mainly to be found in the middle – or male flowers (with stamens only), which are predominantly around the outer edges. Hermaphroditic flowers start off male – the stamens are mature, whereas the pistils aren't – and then they become female when the stamens no longer have any pollen but the pistils are starting to become receptive. This means that self-pollination on a single flower just isn't possible.

As a rule, flowers on the same plant bearing a number of umbels won't reach maturity at the same time, so at any given time the inflorescence will have a flower ratio that is predominantly male or one that is predominantly female. While a large inflorescence remains in flower, female and male phases alternate, thus guaranteeing that an optimum number of tiny flowers are able to be cross-fertilised across the plant as a whole. In effect, the carrot has laid on an open bar.

CAMPANULA
Single bells

Little blue or violet bells discreetly brightening up our country lanes, banks and meadows in springtime: these are campanulas, or bellflowers. Many showier varieties have been selected over the years and are cultivated in gardens. Yet they still retain their unique campanula identity with their little bell-like flowers. The name comes from the Latin, of course, and means simply 'bell', as you might expect. Its bell-like shape is formed by five petals, which are fused at the base and open at the opposite end to produce five little points. This protective petal casing, with the flower facing the ground, protects the reproductive organs, especially pollen, from adverse weather. Yet even before the flowers open into beautiful bells ready to welcome bees or bumblebees on the hunt for nectar or pollen, campanulas remain in bud for an unusually long time for a flower. The bud is shaped like a rugby ball and the very fact that this stage lasts so long should suggest that something interesting is going on behind the scenes.

When it's in the bud phase, a flower is still immature in most cases: its pollen-bearing stamens aren't fully formed, the pistil isn't functional either, nectar hasn't yet been produced, and the petals usually haven't developed their full colour range. If you look closely at a campanula bud, however, you'll see something rather strange happening: the pistil (made up of the style and non-receptive stigmas) is in full growth mode and passes through the middle of the stamens, which are gathered around the centre of the bud and already mature. As it forces its way through, the full length of the style becomes laden with pollen, which sticks to the microscopic hairs on its circumference. When

SCIENTIFIC NAME
Campanula sp.

FAMILY
Campanulaceae

HABITAT
Wasteland, meadows, field margins, upland regions

WHERE TO SEE
Ornamental varieties of various sizes can be seen in gardens, some, like wall bellflowers or Canterbury bells, forming carpets of colour, while other varieties such as Campanula takesimana or Campanula latifolia, the giant bellflower, have much larger flowers.

FLOWERING SEASON
May-July

STRATAGEM

The bell-shaped flowers have a stigma that resembles a star. On reaching maturity, the stigma comes into contact with an insect as it nose-dives into the flower. If the insect happens to be carrying pollen, this will be deposited on the stigma and Bob's your uncle: fertilisation accomplished! If the flower hasn't been visited by an insect by the end of its flowering life, the stigma continues to grow, curling back on itself until it reaches its own pollen and self-fertilisation takes place.

the flower opens, the pistil still isn't fully functional, but it already has a part to play in providing a pollen-rich sleeve ready for any passing pollinator.

The stamens, on the other hand, dangle at the bottom of the bell, virtually devoid of their pollen. This phenomenon is known as secondary pollen presentation. In this kind of arrangement, the campanula flower is male when it first starts to open: the only reward it offers pollinators is pollen.

A few hours later, as the flower ages, the pistil starts doing its stuff: the end of the style splits into three sections and these start to curl back through 90 degrees in three different directions, soon forming three bars at right angles to the petals. Thus forearmed, the stigma, now in three parts, acts as a barrier to any other pollinators that might come along and try to gain access to the nether reaches of the flower. No matter how good a contortionist the insect might be, it's most unlikely that it will succeed in getting past any of the three stigma barriers without touching them. If it happens to have visited a younger flower beforehand, the insect might already be smothered in pollen, which will then be collected by the stigma, allowing successful cross-fertilisation to take place.

But it's also quite possible that no insects will visit the flower. Time goes by, the biological clock is ticking, but still no pollen to fertilise our poor campanula… There must surely be a way around this? Ah yes, fancy that: there's still some pollen stuck to the bottom of the pistils. The very same pollen that was daubed along the style when the flower was still but a bud! But how to access it? Of course, it would be preferable to use pollen from another flower, but when flowering is nearly over, there's not much chance of that happening, so self-fertilisation switches into gear. This is the point at which the stigmas, which have already unfurled to form three bars, but paused there for a moment, continue to develop. They begin to literally unwind all the way down the style until they finally come into contact with a grain of pollen, which fertilises the stigmas and the job is done: the campanula flower has been fertilised. Note that this is a case of 'needs must' and it shows: the fruit resulting from self-fertilisation produces fewer seeds and the next generation of plants will be less vigorous.

From top to bottom

The bell-like blooms of the campanula protect the reproductive organs from adverse weather.

When the stigma reaches maturity, it splits into three separate parts, then curls back on itself at the end of the flower's life.

Botanical plate, 1895

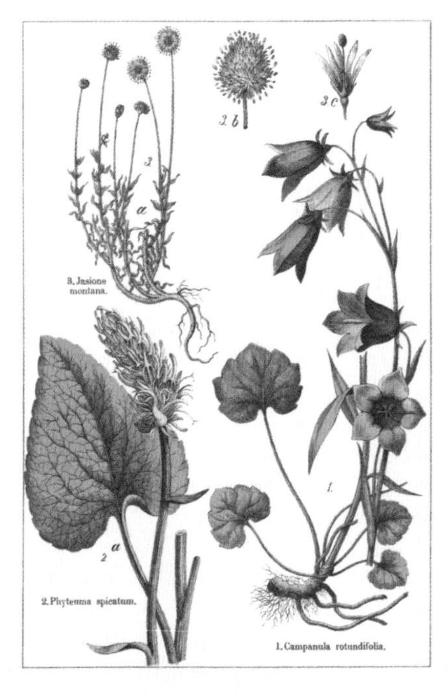
3. Jasione montana.
3 b
3 c
3
a
2 a
1.
2. Phyteuma spicatum.
1. Campanula rotundifolia.

CAMPION
Dancing in the wind

Campions are widespread and can be found in habitats as diverse as forest margins. coastal cliffs and even gardens. These plants rely on various pollination methods, all of which involve insects. However, they also have one other thing in common: they all have a swelling at the base of the flower, created by the fused calyx (collective term for the sepals). This swelling was responsible for their Latin name, *Silene*, in reference to Silenus, a satyr in Greco-Roman culture, whose name was subsequently adopted in Renaissance times and beyond to describe someone who enjoys the good things in life. In Greek mythology, Silenus was the foster father and tutor of Dionysus – he certainly taught him all he knew about wine! In ancient literature, Silenus is described as a hairy man with a squat, dark red nose and scarlet cheeks edged by his beard, heaving his bulk through the villages of mortal souls on a donkey. His legendary drunkenness is said to have involved him in no end of escapades. His pot belly, not unlike an old-fashioned wineskin, might well have been the inspiration for these little flowers, which also have a swollen midriff.

It appears that this anatomical trait, shared by all campion species, is a defence mechanism to protect against the thugs of the pollination world. These include bumblebees and carpenter bees – they don't think twice about hacking into the base of the flowers in search of nectar because their tongues are just too short to reach it by any other means. They gnaw through the flower, making a hole through which they extract the nectar, and leaving the flower unable to attract other pollinators or to take part in the great

SCIENTIFIC NAME
Silene sp.

FAMILY
Caryophyllaceae

HABITAT
Roadside verges, rocky outcrops, coastal areas, woodlands

WHERE TO SEE
When walking in the countryside, on coastal paths or in urban settings. There are a number of ornamental species that are cultivated in gardens, such as the moss campion, Silene acaulis, which is used as a rock garden plant.

FLOWERING SEASON
May-July

STRATAGEM

To protect against bumblebees, which have a tendency to steal the nectar by making a hole at the bottom of the flower, campions have petals that are fused together to form a swelling. The flowers are borne on fine stalks, nodding in the wind in a rhythm easily identified by the insects they are trying to attract. This makes it easier for pollinators to find them.

pollination race.

But when they come up against campions, their usual tactic comes to nothing: try as they might to drill a hole in the bulge at the base of the flower, they are unable to reach the nectar.

Campions are mainly pollinated by butterflies and moths: many have white flowers, requiring the least effort to produce and easily spotted by night-flying insects, but, like honeysuckle, they also release a sweet scent at dusk, attracting moths such as hawk moths (as in the case of the bladder campion, or *Silene vulgaris*).

To prevent self-fertilisation, many campion species separate the sexes with male and female flowers on two different plants.

One small campion species piqued the interest of Welsh scientists when they spotted its special secret to survival in the competitive world of pollination: this is the maritime species, or sea campion (*Silene maritima*). They discovered that the wind makes the campion flowers move in a particular rhythm, which in turn attracts a number of pollinating insects. The flowers are supported by dainty, narrow stalks (petioles), neither too short, nor too long, and dance in the wind with a motion that tempts insects to come their way, almost as if they are waving a flag to attract attention. This movement can be sustained when the pollinator lands on the flower, which means the insects can stay long enough to enjoy both nectar and pollen, and the flower is able to ensure that a few grains of pollen are transferred, albeit by passive means, to the insect's body. The insect in question (the European hoverfly, for example) then flies off, espies another group of white flowers nodding in the sea spray, touches down and extracts the nectar using its proboscis, depositing a couple of grains of pollen on a different sea campion flower in the process.

From top to bottom

The campion appears to protect its nectar from pillagers like bumblebees by means of its swollen calyx.

Drunken Silenus, supported by Satyrs, Peter Paul Rubens, c. 1620.

Botanical plate, 19th century.

However, the scientists also established that if the flowers had an excessively short petiole, limiting their movement, this wasn't as attractive to pollinators, and likewise, if the petiole was too long, this wasn't as efficient either because the flower moves too much and insects are unable to pollinate it effectively.

And thus ends the story of how the little white campion optimises its reproduction opportunities by dancing in the wind beside the deep blue sea.

TULIP

A passionate love affair

Along with the bicycle, tulips are undeniably the jewel in Holland's crown: fields full of tulips in picture-postcard landscapes with canals and windmills in the background. The tulip's heyday arrived when it became a commodity in the days of tulip speculation, making and breaking the fortunes of any number of horticulturists and investors. In essence, this laid the foundation for a liberal capitalist society, where the motive is to make a profit. The tulip mania that engulfed the Netherlands in the 17th century saw the price of tulip bulbs soar to dizzying heights before plummeting back down again. At the peak of this speculative craze, the prices agreed for some bulbs were more than ten years' salary for the skilled craftsman responsible for selecting the varieties. But then the market collapsed, probably when European interest in this new ornamental plant finally waned. Historians reckon that this was one of the first ever speculative bubbles. The highest price ever recorded was for a variety known as *Semper Augustus* – it sold for more than the purchase price of a grand mansion!

And yet this love affair between humans and tulips, which still persists today, is surely down to the fact that this striking spring flower, which brightens up our garden beds and window boxes and forms the heart of so many beautiful bouquets, is so visible. Selection and hybridisation over the years have resulted in thousands of different cultivars. Horticultural tulips were originally developed from a few wild species that grow in the vast plains and craggy foothills of the Middle East and Central Europe; there's even a tulip, albeit stylised, on the Iranian flag.

SCIENTIFIC NAME
Tulipa sp.

FAMILY
Liliaceae

HABITAT
Open position, alpine meadows or steppes, but also woodlands

WHERE TO SEE
In most gardens and on balconies in the springtime, also as cut flowers. There are many national collections in a bid to preserve wild tulip species. The National Botanical Conservatory in Brest is a shining example. Don't miss Keukenhof in the Netherlands, flower gardens extraordinaire and the international tulip capital.

FLOWERING SEASON
March-May

STRATAGEM

The tulip flower is shaped like a goblet with a darker, warmer centre. Nectar is produced in the flower's central recess, a cosy hiding place for insects, as indicated by the individual colours. If tulips aren't pollinated by insects, they can also deploy self-fertilisation techniques.

This large flower is also highly visible to pollinating insects and, in its natural habitat, as in the case of *Tulipa agenensis*, one of the only wild species found in France, there are many elements that favour cross-pollination. However, as we shall see, self-pollination is also an option should pollen-carrying insects fail to drop by.

The tulip is a large goblet-shaped flower with six large tepals (three sepals and three coloured petals, which look similar). In the middle there is a prominent style topped off with a stigma in three separate parts and surrounded by six stamens that produce large amounts of pollen, often black. The flower itself is simple and able to accommodate a whole range of pollinators in its depths. These vary from the cleverest specimens, like bumblebees or honeybees, to the less adept, such as beetles. Deep inside the flower, behind the stamens and beneath the style, pollinators merely need to stick out their tongue to find nectar. As they do this, they become daubed with pollen, which they will go on to deposit, quite unknowingly, on the active stigmas of a different tulip, encouraging cross-fertilisation in the process. However, if there have been no visitors to deposit third-party pollen on the stigmas as the flower reaches the end of its life-cycle, these stamens will tend to bend over towards the middle of the flower, depositing pollen on the stigma and leading to self-fertilisation: better than nothing if fate hasn't been kind to that particular flower.

Nonetheless, it won't be for the want of trying to make themselves look as attractive as possible to pollinators: tulips know how to turn up the heat! First of all, like a number of other flowers, including ipomoeas, tulips open during daylight hours and close at night. The petals respond to heat and open to their maximum when the sun is at its highest. The incident rays of sunlight reflect on the walls of the flower goblet and warm up the centre, just like a solar oven. This phenomenon is even more pronounced when the tulips have a dark centre; in this case, the middle of the flower warms up, which is beneficial for pollinators as winter draws to a close.

From top to bottom

The tulip's dark heart retains heat, which is captured due to the characteristic shape of the corolla.

A small Hymenoptera gathers pollen from a tulip.

Semper Augustus in a vase, Jan Philip van Thielen, 17th century.

Next page (double spread)

Aerial view of tulip fields in the Netherlands.

HONEYSUCKLE

The sweet scent of an evening

Honeysuckle is a nocturnal plant and its flowers are geared towards the pleasures of the night, when it needs to attract the attention of hawk moths and other nocturnal moths that might pass by. No wonder then that the honeysuckle begins to announce its presence, raise its profile and, most importantly of all, release its scent as dusk falls.

The plant's English name, 'honeysuckle', derives from the edible sweet nectar produced in its tubular flowers. The honeysuckle is a climber with an annoying tendency to scramble over anything in its path, its vines weaving through woodland margins or gardens, where it is planted for its spring flowers. The woodland species is pretty tough, producing a great many vines from the same rootstock, which criss-cross their way untidily through surrounding plants with the ability to suffocate saplings unfortunate enough to get in their way. The vines cling to their supports (be they walls, stems, trees or posts) by winding around the underlying structures (not for nothing are they known as twining climbers). Interestingly, all members of the species twine clockwise, although this is pretty unusual in the natural world.

At the top of the long stems, where the vines emerge from hedgerows and woodland margins, clusters of white flowers appear in springtime. These flowers are creamy white, or even pale pink in some cases, and might seem rather dull in themselves in the absence of bright colours. However, this is a defining charateristic of flowers that are pollinated in the evening or at night. White is still visible in the moonlight as it stands out from the dark foliage all around.

SCIENTIFIC NAME
Lonicera caprifolium

FAMILY
Caprifoliaceae

HABITAT
Hedgerows, woodland margins, copses

WHERE TO SEE
They are cultivated in gardens, but also as scented climbing plants clambering over walls and porches.

FLOWERING SEASON
June-September

STRATAGEM

The flowers are highly scented at dusk to attract insects that come out at twilight or in the depths of the night. Their white blooms make them visible in the dark too. They produce nectar at the bottom of a deep tube, perfect for a moth's long proboscis.

The honeysuckle has its sights set on a handful of nocturnal insects and uses both its white flowers and its fragrance at specific times of day to attract them. The scent molecules and nectar are only released at the end of the day when most hawk moths and other night-flying moths start to become active.

Surely everyone knows the heavenly, yet subtle scent of honeysuckle? You can be walking in an enclosed garden on a fine May evening and whoosh, there it is – that glorious honeysuckle perfume. Of course, this fragrance isn't intended for us – its aim is quite selfish: to attract suitable pollinators at the ideal time to allow cross-fertilisation.

Human greed has led to the urge to capture these essential oils and bottle them in perfume form, but honeysuckle doesn't give away its secrets quite so easily. Like many flowers, the very essence of honeysuckle is said to be 'silent'; it is hard to extract it successfully to reproduce the exquisite fragrance of this plant. Its scent has to be reproduced artificially in a laboratory, so any honeysuckle notes you might detect in perfume bottles are the result of mixing synthetic molecules courtesy of the petrochemical industry.

Nectar is produced right at the bottom of a long tube, requiring insects to extend their proboscis as far as possible. This makes it impossible for passing honeybees or bumblebees to access that sought-after nectar: their tongues are simply too short. Of course, some large bees like bumblebees and carpenter bees are devious enough to drill through the base of the corolla to collect their prize that way. But these tricksters are not playing the pollination game because their cheating ploys allow them to bypass the stamens and the pistil.

The honeysuckle's preferred pollinators, on the other hand, all have a very long proboscis and the ability to hover like hummingbirds. The broad-bordered bee hawk moth (*Hemaris fuciformis*), is a good example: the caterpillar's host plants are almost exclusively honeysuckles. While seeking nectar deep within the flower, the hawk moth's hairy body comes into contact with the stamens and pistil, becoming coated with pollen in the process, and this remains stuck fast to its fur until it drops off onto the pistil of another flower. To encourage genetic mixing, the honeysuckle uses a common technique to avoid self-fertilisation: the pistil matures several days before the stamens and ceases to be receptive as soon as the flower's stamens reach maturity. As a result of this delay, the pollen is unable to fertilise the pistil on the same flower. The broad-bordered bee hawk moth is also one of the rare hawk moth species to be out and about during the daytime, but it comes to forage for nectar on the flowers at dusk.

By waiting until nightfall to gather nectar from the fragrant clusters of blooms, this species of moth ensures that these flowers are pollinated.

ROSE

A box of delights

Roses are regarded as the queen of garden flowers, ever the florist's favourite and certainly the preferred option for Valentine's Day. Human beings are so infatuated with roses that they have created thousands of horticultural hybrids from just a few wild species. Their names pay homage to a huge range of famous people including Cardinal de Richelieu, Marilyn Monroe, Brigitte Bardot, Christian Dior, William Shakespeare, Dolly Parton or Lady Diana...

Yet despite desiring these blooms for their beauty and their exquisite fragrance, people tend to overlook one simple fact: these flowers were not originally intended for our enjoyment. The visual and olfactory signals sent out by a rose flower are directed at pollinating insects. And don't forget that the rose, just like any other flower, is a sexual organ.

Rose flowers, especially the kind that have evolved naturally in nature, like the dog rose (*Rosa canina*) or the Apothecary's rose (*Rosa gallica*), rely on a number of key elements to attract pollinating insects and encourage cross-fertilisation.

First of all, these blooms use scent to advertise their whereabouts. Their fragrance is most intense in the morning and is made up of hundreds of chemical molecules that insects like bees or bumblebees pick up via their antennae. As they are drawn in by a fragrant rosebush, the insects' compound eyes start to focus on the flowers themselves. Then, as they get even closer, they are able to make out a large cup-shaped flower made up of five petals in the most simple cases (like the dog rose), or dozens of petals in some hybrid rose varieties. With colours ranging from pink and red to

SCIENTIFIC NAME
Rosa sp.

FAMILY
Rosaceae

HABITAT
Hedgerows, roadside verges, copses

WHERE TO SEE
Everywhere! The rose is the queen of garden flowers and the florist's favourite. There are many rose gardens in France, notably the famous Roseraie de L'Haÿ, just south of Paris.

FLOWERING SEASON
June- November

STRATAGEM

Roses produce a sweet scent and vast amounts of pollen in a bid to attract insects. After years of breeding, some roses produce very little pollen and what they do produce is virtually inaccessible. Others have also lost their fragrance, making them less attractive to pollinators.

white or even yellow, the blooms contrast with the green of their foliage.

Hundreds of tiny yellow hairs glisten in the centre of the corolla: these are the pollen-bearing stamens. This pollen isn't just the flower's male sexual organ but also, along with nectar, the main source of food for pollinating insects. There is a great deal of pollen, which is just as well because that's all roses have to offer their bee visitors. Roses don't produce nectar – so do be careful if you happen to see a pot of honey labelled 'rose honey'. This abundance of pollen attracts not only bees and bumblebees, eager to collect their booty in little pouches wedged between the hairs of their hind legs and carry it back to the colony to feed their larvae, but also beetles, some of which binge to excess and even eat the petals as well. However, these visitors, be they bees or beetles, almost always find themselves covered in pollen. Some remains in situ while the insect flies off and lands on another rose flower of the same species. That's when the pollen is able to come into contact with the stigmas at the tip of the pistil, at the heart of the crown of stamens. Pollen from the first flower passes down the pistil until it reaches the ovule, where it deposits a sperm cell: cross-fertilisation accomplished.

And yet humans, with their annoying tendency to monopolise natural resources, have managed to modify the original natural roses by successive hybridisation and selection techniques to produce up to 30,000 different eye-catching varieties. In some cases, the selection process is based on reinforcing mutations that have come about naturally, such as when stamens have developed into petals, resulting in roses with large numbers of petals (old roses like *Rosa Cuisse de nymphe émue*, for example), but with fewer stamens, effectively reducing the food on offer for visiting insects. By the same token, modern varieties are often scentless; successive selection processes have bred out the scent genes used to produce the odour molecules. All of which means that modern roses may be superb specimens to the human eye but deprive rose bushes of both of their seductive attributes from the pollinators' perspective: perfume and pollen.

From top to bottom

The English rose Strawberry Hill has the look of an old rose, with a heady scent and many petals, which originally evolved from stamens. Insects struggle to access food in these kinds of roses.

A modern red rose is the consumer society's symbol par excellence of human love, yet it often has no fragrance or stamens and is therefore of no interest to pollinators.

Roses in a stemmed glass, Henri Fantin-Latour, 1873.

CALLISTEMON

Plenty to go around!

With its soft tufts of long red hairs, the callistemon is not a typical flower. In fact, its brush-like form has earned it the common name 'bottlebrush'. Callistemons are originally from Australia and come from the same family as the eucalyptus (Myrtaceae). If you look closely at one of its so-called flowers, you will realise that each long-haired spike is in fact a small bunch of long red stamens with sacs of purple or yellow pollen, each belonging to a single flower, which doesn't have any petals. The pistil is located at the heart of the plume of stamens with plentiful nectar resources at its base. In other words, a bottlebrush is made up of nearly a hundred tiny flowers arranged in a sleeve-like formation around the shrub's stem.

Australia is home to some extraordinary animals that have evolved in a unique and isolated environment over a period covering millions of years. The bottlebrush is equally extraordinary and some of its pollinators are also not what you might expect. Take, for example, the tiny marsupial that climbs up the shrub and then, like a tightrope walker, balances at the bottom of the brush, with its front paws moving the leaves to one side and grasping the stamens, while its long tail curls around the stem. Meet the eastern pygmy possum (*Cercartetus nanus*), otherwise known as the honey possum, which pokes its pointed snout into the plume of flowers, getting its fur smeared with pollen in the process. It then uses its long tongue to guzzle as much nectar and gather as much pollen as it possibly can. Its teeth are fused together to form a kind of comb that enables it to collect and then chew the precious pollen as it passes its pollen-laden tongue over the comb. The honey possum,

SCIENTIFIC NAME
Callistemon sp.

FAMILY
Myrtaceae

HABITAT
Copses, hedgerows, open woodland

WHERE TO SEE
Originally from Australia, the callistemon or bottlebrush is available in a variety of different-coloured cultivars. Some varieties are cultivated in France, especially in the south or in coastal areas where there is no risk of frost.

FLOWERING SEASON
March-July

STRATAGEM

The flowers look like bottlebrushes, with many long stamens that transfer pollen to the coats of a number of rather surprising animals: not only marsupials, which have evolved to eat only nectar and pollen, but also birds and bats.

which feeds exclusively on nectar and pollen, is nocturnal and locates these flowers primarily by their scent.

During daylight hours, the bold red colours of the callistemon send out strong signals to the many nectarivorous birds found in Australia, which can justifiably describe itself as a paradise on earth for wildlife. At daybreak, the shrub often plays host to a multicoloured flock of happily chattering visitors: rainbow lorikeets feasting on all the nectar they can find before moving on to another bush.

Then, at twilight, other flying creatures alight on the callistemons to take their fill of the pollen and nectar bounty – bats! These are excellent pollinators and Australia has a great many species, some of which have a wingspan measuring as much as a metre and a half. One example is the grey-headed flying fox, an iconic tropical bat, which occasionally feeds on nectar from the callistemon.

Like it or not, all these unusual pollinators end up with their fur or plumage smothered in pollen as they hunt for nectar at the base of the flowers. This is all down to the ingenious design of the callistemon: the very structure of the flower, with its long stamens that bend and move apart when they come into contact with the body of a pollinator on the quest for nectar, is perfectly suited to dispersing their pollen. The bottlebrush works rather like a make-up brush, the kind you might use to apply face powder, gently dusting the visiting animals' fur or downy feathers with a soft cloud of yellow or purple pollen.

Each inflorescence has a great many stamens for dusting purposes; prospective pollinators will never be able to remove every single grain of pollen, no matter how often they groom themselves. This makes it highly likely that one of these grains will find its way to another flower, on a different bush, when creatures like the honey possum turn up ready for another banquet. Inevitably, they touch the pistil located in the middle of the large stamens and before you know it, the callistemon has been cross-pollinated.

HIBISCUS
Right or left?

The hibiscus, symbol of many a holiday on a tropical island with white, sandy beaches, surfboards and hummingbirds, is an exotic plant par excellence. And rightly so: many hibiscus species do indeed come from the tropics, but here we are particularly concerned with the renowned Hawaiian hibiscus. Native not only to the Pacific, but also to South East Asia, this hibiscus is also the national flower of Malaysia. It is reputed to have been imported to Europe from China, hence its common name, Chinese hibiscus, from the Latin *Hibiscus rosa-sinensis.*

This hibiscus is red – in-your-face red, in fact! The large scarlet flowers readily stand out against the bush's dark green leaves and the surrounding tropical vegetation. You can certainly see them from afar... However, their bright red colour isn't always perceived as red: insects, particularly bees, don't have the ability to identify red and would struggle to distinguish the flower from the leaf. That said, birds, especially hummingbirds, do appreciate the breathtaking beauty of the Chinese hibiscus. As for scent, even if you stick your nose right inside the petals, you won't have any joy. Like all flowers that are primarily pollinated by birds, hibiscus don't waste their energy producing scent molecules because birds have a very limited sense of smell. But hummingbirds should be grateful they aren't nocturnal, as hibiscus flowers close up at dusk to protect their nectar and pollen from night-foraging herbivores.

In some hibiscus species, to guarantee cross-fertilisation, the pollen cannot be deposited on the pistil of the same plant thanks to the long central stalk that carries the male and female pollination organs, as in

SCIENTIFIC NAME
Hibiscus rosa-sinensis

FAMILY
Malvaceae

HABITAT
Groves, hedges

WHERE TO SEE
This species won't tolerate frost but can be cultivated in pots or grown in mild climates, especially in coastal areas. Other hibiscus species, such as Hibiscus syriacus, are more hardy and are often grown as hedging plants.

FLOWERING SEASON
March-October

STRATAGEM

The plant has large, red, scentless flowers, as is typical of flowers pollinated by birds. Some hibiscus species have flowers in which the pistil.points to the right, while others have flowers with the pistil pointing to the left. If pollen is gathered from a flower pointing to the right, it can only be deposited on a pistil pointing to the left, thus ensuring cross-fertilisation.

the case of all members of the Malvaceae family (hollyhocks, for example).

The stem that protrudes from the corolla consists of a long style surrounded by a cuff of spherical stamens, much like little pompoms just waiting for a bird to brush past them to relinquish their grains of pollen; at the very end of the style there are between four and five stigmas, also spherical. The body of a pollen-dusted bird merely has to touch the stigmas for some of its load to be deposited on the male gamete – causing the flower to be fertilised as a result.

But how does the flower manage to get its pollen onto the bird's body and then onto the stigmas in the first place? More importantly, how does it make sure that the pollen is from a different plant?

Hibiscus flowers produce a plentiful supply of nectar in nectaries at the bottom of the petals. This food source is much prized by birds like hummingbirds. They insert their long beaks right down inside the flower, hovering in the air and brushing against the stamens in the process. When they fly off, looking for nectar in another hibiscus flower a little further on, their body comes into contact not only with the stamens, but also with the stigmas of the new flower. Their pollen cargo is then deposited on the stigmas.

This is all based on a cunning trick: the styles, with their pollen cuff, don't all have the same configuration. The end of the style, where the stigmas are located, tends to veer either to the left or to the right. This trait has been found to be particularly noticeable in *Hibiscus schizopetalus*.

From top to bottom

A hummingbird feeding from a hibiscus flower with prominent stigmas, but no contact with the pollen.

Close-up of the reproductive organs: hairy, globe-like stigmas, and stamens.

Different stages of a hibiscus flower.

In flowers that veer to the left, hummingbirds in search of nectar will only become coated with pollen on the right of the flower; as they take off to the left, the style prevents any contact between the bird and the pollen located on the left. The reverse happens with bushes that have flowers veering to the right. In other words, pollen is deposited on the opposite side of the stigma. The hummingbird can visit as many other flowers on the same plant as it likes, but the pollen will never come into contact with the stigmas. As it ventures onto a different plant, the pollen deposited on the right-hand side of the hummingbird will come into direct contact with the stigmas on the left of the new flower, leading to successful cross-pollination.

HELLEBORINE

The magician of the woods

As we have observed, orchids are queens of deception, skilled at attracting insects for their own gain to ensure cross-pollination. Some (like the bee orchid or hammer orchid) mimic the female form to attract males in search of a mate, while others, like bucket orchids, use essential oils to entice males.

Epipactis helleborine (or broad-leaved helleborine) has opted for a different stratagem, though no less amazing. This unobtrusive little orchid flowers at the height of summer, deep in the forest, in dark, damp environments. In these kinds of places, where the sun's rays never penetrate the dense foliage, pollinators are few and far between. However, at this time of year, social wasps like the common wasp (*Vespa vulgaris*) – familiar to anyone who's ever tried to eat a ham sandwich or juicy peach on a summer picnic – are out in force and in search of food for their growing larvae back in their paper nests.

Wasp larvae are partial to meat, especially nice, plump caterpillars. The worker wasps in charge of food rations for the entire colony are only too aware of this preference and put these delicacies at the top of their foraging list.

Back to our helleborine though: evolution has done its stuff, and as a result of successive mutations, not to say trial and error, a mutual relationship has developed between the orchid and the caterpillar-seeking wasps. The orchid needs insects to transfer its pollen from one flower to another, while the wasp needs caterpillars to feed its larvae. But so what?

SCIENTIFIC NAME
Epipactis helleborine

FAMILY
Orchidaceae

HABITAT
Herbaceous undergrowth in woodlands on plains or in mountain areas (up to 2,000 m)

WHERE TO SEE
During summer walks or hikes in the forest. Although sometimes found naturally occurring in gardens, like all orchids, the helleborine is very hard to cultivate.

FLOWERING SEASON
June-August

STRATAGEM

The green and brown flowers, barely visible, give off a scent similar to that released by plants when they are infested with caterpillars. Hey presto: wasps in search of caterpillars are drawn to the flower and dive in!

Wouldn't it be a good idea if the orchid could conjure up caterpillars, just below its pollinia? But wait, a plant that produces insect larvae – surely not? Nature always has the capacity to surprise us, but this sounds much too far-fetched to be true. And no, the orchid doesn't go quite this far – instead, our little helleborine flowers release the same kind of odorous substances that the leaves of other plants usually give off when they are being attacked by caterpillars. If you're a plant, it must be pretty annoying if caterpillars come and devour your leaves. So, if it's unable to move, a plant needs to find a solution to handle such attacks. They use a range of strategies to defend against herbivores (enough material here for another book, perhaps?), one of which entails producing specific scents to attract the kind of insects that eat caterpillars. Self-defence, in other words.

The orchid uses this olfactory signal to its advantage. The flower isn't actually being attacked by caterpillars at all, but, by pretending that it is, the orchid is able to attract wasps desperate to find caterpillars to take back to their colony.

On arriving at the source of the would-be banquet, they are faced with just a flower, with stamens and pollen – oh, and nectar on the side. But now they are there, the wasps might as well indulge themselves. The orchids aren't heartless enough to carry on the deception, so they provide copious amounts of nectar, meaning that at least the wasps don't go away empty-handed. The nectar they supply seems to make the wasps groggy; no matter how active they were beforehand, they become much more ponderous, hesitant and clumsy after imbibing. This is actually no bad thing for the orchid; the wasps spend more time tottering around the flowers and picking up the pollinia (those pollen-stuffed globules) on the top of their heads, not realising that this pollen will end up being transferred to the pistil of another flower. So what's the secret of this special nectar? It contains a powerful narcotic: oxycodone. This substance is a member of the opioid family, often manufactured synthetically and used medicinally in humans; it is also said to be extremely addictive.

STINKING HELLEBORE

Winter warmth

In midwinter, when much of the natural world is dormant, one little woodland flower seems able to withstand the cold. The stinking hellebore proudly displays its pale green, bell-like flowers in clusters a couple of feet off the ground and well above its dark green, deeply divided leaves. As a bonus, in temperate woodlands, its flowers are in full bloom at a time when other flowers are thin on the ground, thereby reducing competition. Unfortunately, there are very few insects around at this time of year: honeybees are tucked up nice and warm in their hives and many adult solitary bees have already died off.

Bumblebees, like the buff-tailed bumblebee, have the ability to vibrate to raise their body temperature and also have a very thick, furry coat: they are still active late in the season until the new queens hibernate. Nevertheless, these young queens do sometimes emerge from hibernation very early, even when the thermometer is as low as 4 or 5°C, as long as there's a little bit of sunshine to warm up the ground. This is why the bumblebee is the chosen candidate for pollinating hellebores, be it at the beginning or end of the winter season. And just as it isn't unusual to see bumblebees out and about even when the forest floor is carpeted with snow, it's not uncommon to spot hellebore flowers peeping out of a snowdrift: the Christmas rose (*Helleborus niger*) is a case in point.

Take a good look at stinking hellebore flowers: working inwards, you'll see five pistachio-green sepals forming the calyx, sometimes tinged with red at the tip, and then an array of stamens encircling three to five pistils. But where are the petals? If you peer between the stamens and the sepals, you'll find some

SCIENTIFIC NAME
Helleborus foetidus

FAMILY
Ranunculaceae

HABITAT
Woodlands, scrubland and rocky soil. Found at altitudes of up to 1,800 m

WHERE TO SEE
In forests or gardens in the depths of winter; many hellebore species are horticultural cultivars and widely available in garden centres.

FLOWERING SEASON
January-April

STRATAGEM

The flowers open in midwinter, their petals transformed into highly productive nectar tubes. Nectar contains yeast, which causes the sugar to ferment, and this process releases heat, which attracts pollinators, especially bumblebees.

very special nectaries: the petals have been modified to create little tube-like structures supplying a veritable cornucopia of nectar.

It's as though the hellebore has pulled out all the stops: shy and delicate it may be, but this drooping flower (a trait that protects its wonderful storecupboard from rain and hail) is keen to treat those rare pollinators willing to assist in perpetuating the species with an impressive quantity of pollen and nectar. And, just like any other restaurant worthy of the name, the plant makes sure its marketing messages are right up-to-date: as soon as the larder is bare, the hellebore lets punters know – a red border appears at the top of the sepals, like a warning sign indicating that this flower has nothing more to offer.

As in many flowers, but also like a good village bakery, scent is another signal they often use. Hellebores have a unique mechanism for releasing volatile odour compounds into a very cool atmosphere, which is not really conducive to diffusing scents: nectar contains a high concentration of yeast, which, as it metabolises the sugars in the nectar, generates heat. In turn, this heat helps to disperse the volatile odour compounds and increases the temperature in the heart of the flower to 2°C above ambient temperature. Studies have shown that, if they have the choice, bumblebees tend to prefer the warmest flowers. Given that, to the bumblebee, the flower must feel like a five-star hotel in the depths of winter, there's no doubt that they enjoy the microclimate in the centre of the flower. As well as offering board and lodging, the hellebore also provides a cosy dining room. However, these bees also – perhaps most of all – enjoy a sugar-rich, slightly alcoholic nectar. As it produces heat, the yeast activates the fermentation process (like when making beer or wine), consumes the sugar and produces alcohol. In other words, this is a symbiotic relationship with three partners: the flower, the insect and the yeast – it's all a question of balance.

From top to bottom

A honeybee (*Apis mellifera*), laden with pollen, feasting on nectar in a hellebore flower.

The hellebore is made up of a number of petals that have been converted into nectaries, forming long, dark green horns of plenty in the heart of the flower.

Botanical plate, 20th century.

4
2
1
6
5
3

ARUM
A very particular smell

What a strange flower, with one large petal surrounding an erect, skywards-pointing spike. But its pollination mechanism is even stranger still. When you first come across a speckled arum flower in a damp and shady corner of the forest, you can't help but notice its unusual shape: a large, almost conical, greeni-sh-white mainsail – this is the spathe. The spadix, a red or yellowish club-like structure, rises proudly from the centre of the spathe. You might think that this carries the pollen, in full view of passing pollinators – but you'd be wrong (or at least not entirely correct).

In fact, to ascertain exactly what's going on inside this flower, you need to look beneath the petals or, in this case, beneath the spathe. If you move the spathe to one side, you can see the bottom of the spadix, where there are four clearly defined areas. First of all, there are a hundred or so little white buds – these are the female organs: hundreds of pistils emerging from ovaries, which, once fertilised, will turn into fruits. Above these is an initial barrier comprising long hairs poin-ting downwards. Next, there is a ring of male flowers, which, when mature, will produce pollen. Finally, just above the male flowers, there are yet more long hairs, forming a second barrier.

But who on earth will be willing to make their way down into the depths of these flowers to pollinate them? Bear in mind, too, that arums aren't quite as generous as other flowers: they don't even produce any nectar! T hey really have nothing to offer.

Like many orchid species, arums depend on the vital needs of insects to exploit their pollinators'

SCIENTIFIC NAME
Arum italicum

FAMILY
Araceae

HABITAT
Damp and shady woodland understorey

WHERE TO SEE
Wild arums are usually found in woodland settings. Many ornamental varieties are available from garden centres and florists. The titan arum is the world's largest species and certainly makes a statement in a number of botanical gardens, including those in Brest and New York.

FLOWERING SEASON
April to June

STRATAGEM

The flowers are made up of a large petal surrounding a club-like structure and forming a constriction at the base of the flowers. They attract flies by emitting strong odours which are reminiscent of the droppings where flies like to lay their eggs. The flies become trapped in the constricted area and emerge covered in pollen, ready to fertilise another arum.

mobility and transport their pollen. They play to the reproductive instincts of certain insects, namely flies and gnats, which tend to lay their eggs in the least desirable places: animal dung or rotting flesh (as in the case of the dead horse arum lily, found in Corsica and Sardinia).

A very specific target audience, in other words. As to the mechanism they use, it's even more Machiavellian: the arums set a trap. They use hairs to keep insects captive for as long as possible so they are forced to gather or deposit pollen.

But how do they attract flies into their trap when these insects are known to lay their eggs in animal droppings? As is often the case, the best way of doing this is by recreating the scent that most appeals to them. Arums use volatile substances to conjure up the distinctive smell of urine, excrement and rotting flesh. To make this odour even more realistic and provide a cosy environment for their prisoners, arums have the ability to generate heat at the bottom of the spadix, sometimes reaching as much as 15°C above ambient temperature. This added warmth also helps the delightful aroma to drift even further afield. In one infamous species, the titan arum, which has the world's largest flower and is up to three metres tall, the enormous spadix, armed with its own central heating system, is able to disperse its unique fragrance within a radius of up to eight hundred metres.

Once flies have been lured to the centre of the spathe, they progress down the spadix and pass through both sets of hair barriers (designed to allow one-way traffic only). When arums start to flower, only the female organs are mature; pollen-laden flies arriving from other, older flowers are able to deposit their cargo on the flowers, thus ensuring pollination. The flies are kept captive at the bottom of the flower for several hours, if not a whole day. The female flowers then stop being receptive just as the first barrier of hairs starts to wither away, clearing the way to the male organs, which have now reached maturity. The flies become coated in more pollen as they go past, then are able to exit through the second barrier, which soon starts to wither in its turn. Bear in mind that the flies haven't been able to feed or lay eggs during their time of captivity, although this doesn't seem to stop them being tricked next time round. They are still convinced that they are going to be able to lay their eggs on a tempting piece of rotting flesh – only to find that it's another arum flower. When they reach the base of the flower, they are once again trapped in by the hairs, but as they try to escape, they transfer their pollen to the new arum flower and contribute to cross-fertilisation in the process.

Next page (double spread)
This titan arum (*Amorphophalus titanum*) is the largest flower in the world; here, flowering in the New York Botanical Garden in 2018.

IVY

Last fuel stop before winter

Clinging onto trees or old walls, ivy is part of our collective imagination, conjuring up images of country cottages or old manor houses. The age-old association between ancient stone and creeping ivy even led to the collective name for eight prestigious universities on the American East Coast, otherwise known as the Ivy League. Age can be a guarantee of quality and their ivy-covered buildings bear witness to their long history.

Ivy is a quirky kind of plant, living its life slightly out of sync with the trees and other plants it grows alongside. Its main growth season is at the end of summer and in autumn, when trees are starting to lose their leaves; this enables the ivy to take advantage of gaps that appear in the foliage canopy, allowing it to spread out its long stems in the pale autumn sunlight. Similarly, its flowers appear well after the intense main flowering seasons of spring and summer. In the northern hemisphere, they open in September or October, often the only blooms around in most of the ecosystems where ivy grows. Once its flowers have been pollinated, they develop into fruits in the depths of winter, just in time to provide a rich food source for both non-migratory birds and mammals that don't hibernate in the colder months. It is their task to disperse the ivy seeds.

Ivy flowers certainly aren't the showiest blooms on the block, to say the least. And yet, at the peak of their flowering season, any number of solitary bees, honeybees or bumblebees, as well as some wasp species, butterflies and flies, can be seen buzzing happily from flower to flower.

So let's look at an ivy flower: it has ridiculously small

SCIENTIFIC NAME
Hedera helix

FAMILY
Araliaceae

HABITAT
Growing on trees, on the ground, in forests and hedgerows

WHERE TO SEE
Commonly found in towns and cities, growing on walls; it is also available in garden centres, especially as ornamental varieties like the variegated specimens.

FLOWERING SEASON
End of September-October

STRATAGEM

Ivy flowers are tiny and almost invisible, with very short petals. However, the fact that they are clustered together in a spherical formation makes them more attractive to insects. They also provide copious amounts of nectar just as winter sets in, when there are very few flowering plants in the woodland understorey.

petals, pale green in colour and almost invisible. Right at the heart of the tiny flower, you can see the pistil, surrounded by a dark mass, from which five stamens emerge.

It certainly doesn't look very appealing. Ivy's primary strategy works on the principle of the more the merrier: it doesn't produce isolated flowers but little, upright, spherical clusters containing hundreds of individual blooms. Clusters are more likely to catch the eye than single flowers, after all. Another of its ploys is to provide enough nectar to feed all those hungry insects still around this late in the season.

The first frosts of winter are just around the corner, yet many pollinators are still active and need more food to survive when the weather turns chilly. For honeybees, ivy is often their last chance to fill up with nectar, which they can turn into honey for the winter months once they return to the hive. In other words, since ivy produces so much nectar, it is a key resource for these insects. And while this abundant and very rich nectar produces excellent honey, no self-respecting beekeeper with an eye to the wellbeing of their bees would dream of selling ivy honey: this remains in the hive to help the colony survive the winter.

Ivy may be out of step with other plants and therefore regarded as an outsider, but although it attracts a great range of pollinators, it also encourages a very specific insect to step out of line too. The ivy bee (*Colletes hederae*) is a species that is almost exclusively dependent on ivy for its survival. It is a solitary bee – in other words, it doesn't live in colonies, nor is it keen on the division of labour: female ivy bees not only reproduce, they also forage for sustenance to feed their offspring.

From top to bottom

Clusters of ivy flowers.

A honeybee (*Apis mellifera*) sticking out its tongue to extract some of the abundant nectar from an ivy flower.

Botanical illustration by Jean-François Turpin, published in *Flore Medicale* in 1830.

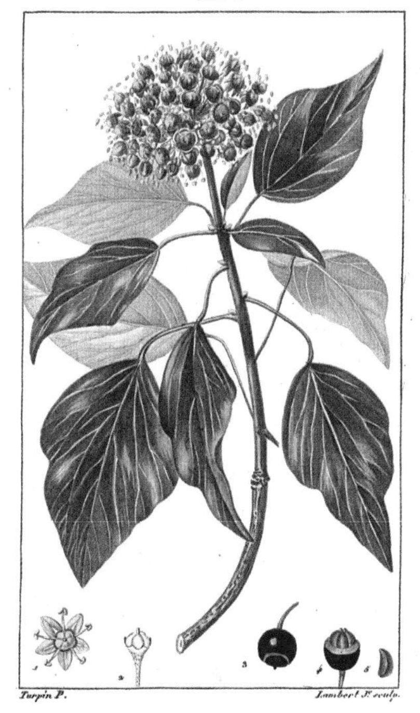

218.
Turpin P.
Lambert J.e sculp.
LIERRE.

HYDRANGEA

This is where it's at

Mophead and lacecap hydrangeas refer to different cultivars, originally from Asia or North America. Lacecaps, with their dark green leaves and inflorescences made up of tiny, unremarkable flowers surrounded by eight to ten larger flowers in shades of pink, mauve or blue, have adapted to grow in all temperate regions and are the undoubted stars of gardens in Brittany or the South of England.

Lacecap varieties have two types of flowers: small ones with tiny petals, which aren't anything much to look at, and larger flowers with brightly coloured petals arranged around the outer edges of the inflorescence. Mophead hydrangeas, on the other hand, have ball-like inflorescences, consisting solely of the same brightly coloured petals you find around the edges of lacecaps.

First, we're going to concentrate on the fascinating pollination story behind lacecap hydrangeas and leave mopheads to one side for the time being. Let's start by inspecting the little flowers in the middle: there may be a lot of them, but they are pretty insignificant. Look closer still and you'll see the flower's entire sexual armoury on show: the pollen-bearing stamens surrounding a pistil. Then there are the nectaries, which supply nectar and attract all kinds of pollinators.

Now let's turn our attention to the flowers around the outer edges: they are anything but insignificant! You can see them from afar, with their four petals in striking colours, BUT they don't have pollen, nor a style, nor any nectar. These flowers are sterile.

So what's the point of the plant producing flowers without any sexual organs? Well, imagine you're a bee

SCIENTIFIC NAME
Hydrangea macrophylla

FAMILY
Hydrangeaceae

HABITAT
Woodland understorey; they like semi-shade and damp conditions

WHERE TO SEE
Hydrangeas are planted throughout Brittany, as well as in coastal or mountainous regions on granite terrain. They form impressive blocks of colour in gardens. There are many hydrangea collections in the forest of Brocéliande in Brittany, or in Rainans in the Jura region of eastern France.

FLOWERING SEASON
June–August

STRATAGEM

The flowers are grouped together in clusters of tiny, insignificant flowers, which produce pollen and nectar, with large, sterile flowers around the outer edges to attract insects to the central flowers. The petals of these outer blooms face skywards, but as soon as their nectar supplies are exhausted, they droop down, letting insects know that the larder is closed.

looking for food. Those sterile pink flowers stand out against the dark green of the hydrangea leaves. With its compound eyes, the bee is able to distinguish the inflorescence and heads straight for the food source.

Ignoring the sterile flowers, which don't have anything to offer, the bee moves on to the tiny flowers in the centre of the flower head. After topping up with nectar, picking up some pollen on its legs and unwittingly getting its body dusted with pollen, the insect flies off again, spots other sterile flowers on a different hydrangea, and uses these as a guide to zoom in on another set of central flowers. The bee continues to gather nectar, not only for its own use, but for its larvae too; as it flits from flower to flower, it deposits precious grains of pollen on the stigmas of these new flowers, fertilising the hydrangea as it goes.

But the hydrangea's ingenious ploys don't stop there. While the sterile flowers point insects in the direction of their next nectar feast, they can also signal quite the opposite. These sterile blooms can also tell bees to keep on going: nothing to see here – this restaurant is closed. Once all the fertile flowers in the middle have been pollinated, they stop producing nectar and are no longer receptive to pollen. No further need for visitors. As soon as pollination is accomplished, this gives the sterile flowers the signal to stop attracting pollinators. Whereas previously their petals were wide open, pointing skywards and proudly on display, they suddenly turn downwards, facing the ground and their colours become more muted. A clear signal that there's nothing left to eat and no need for visitors to drop by.

But what about the other kind of hydrangea: the mopheads, which only have sterile flowers? These cultivars are rarely fertilised. They are usually cultivated by vegetative propagation, by dividing a bush with several stems to create new plants. The sterile flowers don't produce fruit or seeds and their vividly coloured petals gradually fade but remain in situ. However, on very rare occasions there are sometimes a few fertile flowers, which does allow a certain degree of genetic mixing within the population. And whatever pollinators might think, mophead hydrangeas, with their stunning ball-like heads, make beautiful bouquets of dried flowers.

From top to bottom

Kingfisher and Hydrangea,
Utagawa Hiroshige, c. 1830.

White hydrangea on gold paper,
Japanese print, 19th century.

The false flowers around the edges of this hydrangea face the ground, indicating that there's no nectar left in the fertile flowers.

EDELWEISS

Star of the Alps

Majestic peaks, mountaineering and fresh air: that's what springs to mind when you hear the word 'edelweiss', with its little white flowers sparkling like silver treasure. The edelweiss, or 'noble white flower' (as its German name suggests, from *edel,* meaning noble, and *weiss,* meaning white), is an iconic mountain flower, but surprisingly rare, scattered as it is across the steep slopes of the Alps and the Himalayas; it grows at altitudes in excess of 1,300 m.

Indeed, it's the flower's very specific habitat that has made the edelweiss, like a silver star set atop a delicate, leafy stem, so distinctive. Mountains are hostile environments for any flowering plant, posing all kinds of hazards. First of all, temperature: at these altitudes, it can be extremely cold, just as it can also be extremely hot, and temperatures fluctuate in a very short space of time. From night-time to midday on a sunny summer's day, they can range from -10°C to as much as +40°C! Then there's drought: mountain slopes are exposed to wind and they are that much closer to the scorching sun at the height of summer. No wonder the terrain easily dries out during this season. Last but not least, the sun itself: it may be necessary for plants to survive, because it allows them to photosynthesise, but the atmosphere at higher altitudes is thinner, which makes it harder to filter the sun's rays, particularly the ultraviolet variety, which are particularly powerful high up in the mountains. Ultraviolet light has a negative effect on plant tissue.

Just as mountaineers wear specially adapted clothing, sunglasses and sunscreen when they venture onto the peaks, in order to survive in these extreme conditions,

SCIENTIFIC NAME
Leontopodium nivale

FAMILY
Asteraceae

HABITAT
Open rocky outcrops in the mountains at altitudes of between 1,800 and 3,000 m

WHERE TO SEE
On hikes in the Alps, and in botanical gardens in alpine areas.

FLOWERING SEASON
July-September

STRATAGEM

The flower is in fact a collection of tiny flowers arranged in a rosette on top of a star formed by velvety leaves. The flowers are successively male and female, thus permitting cross-fertilisation. A silvery down protects them from the harsh mountain conditions caused by extreme cold, drought and ultraviolet radiation.

the edelweiss has developed its own special protective clothing in the form of its beautiful velvety silver ensemble. This serves various purposes to enable the plant to withstand the challenges of growing at this altitude: its thick down traps in air and insulates the flower, making it less sensitive to huge temperature swings.

The flower's white hairs also prevent evaporation and protect the plant in times of drought. Being white, the petals also reflect light, which stops it overheating. Finally, the hairs protect the plant tissue from ultraviolet rays.

As the flower develops, a number of golden yellow dots start to appear in the fuzzy covering. Take a closer peek and you'll see that the edelweiss is actually an inflorescence. Masses of tiny flowers form right in the middle of what were once thought to be grey petals – these are the flower heads, or capitula, of miniature disc florets: small flowers with their petals fused together to form little tubes. The edelweiss is a member of the same family as sunflowers and thistles: it's an Asteraceae.

When it first starts to flower, a dozen or so large false petals (or bracts), silvery grey in colour and furry to the touch, surround an initial flower head dotted with tiny pollen-bearing flowers. The pistil at the centre of the flowers isn't functional: these are male flowers. Female flowers are arranged around these male flowers: this time, it's the stamens that aren't functional. The male and female flowers are separate, in other words. The first flower head is then followed by a number of other flower heads that gradually reach maturity, covering the flower with many rounded yellow spots.

From top to bottom

Edelweiss, Edgar Maxence, 20th century.

A European wool carder bee (*Anthidium manicatum*), probably about to steal some of the edelweiss's woolly fleece to line its nest.

This edelweiss has a central flower head with the male blooms in full flower.

The pollen from the male flowers, which open first, is unable to be deposited on the female flowers on the same flower head because the female flowers are not yet mature. All of these flowers also produce a concentrated form of nectar, which is particularly rich in amino acids essential for the survival of certain flies; these are in fact some of the few insects to live at these altitudes. As they feast on nectar at the base of the male flowers, they pick up pollen on their legs or between their hairs, which they then deposit on the mature female flowers of a different edelweiss as they move in to devour yet more nectar. All of which means that edelweiss are cross-pollinated by flies with the ability to survive at these altitudes.

Next page (double spread)

Edelweiss grows at high altitudes up to 3,000 metres, such as the Dolomites, as in this picture.

SNAPDRAGON

In the jaws of the dragon

There are some flowers that don't do things by halves: while keen to protect their reproductive organs, they are also determined to increase their chances of pollination. Snapdragons, or antirrhinums, are liable to swallow, dragon-like, any pollinator passing by in search of pollen or nectar. And rest assured, there are many insects hungry enough to try, because the snapdragon is very much in demand. This late-spring flower, often found growing on rocky outcrops, along walls or in gardens, has an attractive fragrance. But when insects come face-to-face with the inflorescence, they may initially be nonplussed: how do they get in? Where's the pollen – and the nectar? They need to follow the signs: the fused petals form a large lip topped by two yellow patches against a pink background. Once in situ, drawn in first by the scent, and then by the colour, insects may feel reassured that they're on the right track. Surely the prized nectar must be close at hand? This is where the fun starts: it looks as though they'll have to get right inside the flower, but they'll have to be quite hefty to gain access, as it's no easy matter. Honeybees, mason bees, butterflies and flies are brought to a halt at the snapdragon's threshold; they are simply too light to open the antirrhinum's access mechanism. Bumblebees, on the other hand, are perfectly suited to the job. Equally, the snapdragon's entrance portal seems perfectly suited to the bumblebee: its lower petal acts as a landing pad, but the bumblebee needs to remain in a vertical position to access the flower. The petal is ideal for climbing, covered as it is with conical cells in exactly the right location to stop the bumblebee from slipping. Once

SCIENTIFIC NAME
Antirrhinum sp.

FAMILY
Plantaginaceae (formerly Scrophulariaceae)

HABITAT
Rocky outcrops in the Mediterranean basin and the high Pyrenees

WHERE TO SEE
When walking on limestone scree in the Mediterranean region, or on the foothills of the Pyrenees. Many ornamental varieties are cultivated in gardens.

FLOWERING SEASON
July-October

STRATAGEM

The flowers are closed and protect the reproductive organs. When a bumblebee lands on a flower, it opens like a hinged jaw, enabling the bee to get right down to the bottom to reach the nectar. The bee then comes into contact with the pollen or the pistil and cross-fertilisation is complete.

inside, it can slide down the corolla to find its heart's desire: the nectar supply at the bottom of the flower.

In most cases, the snapdragon's 'mouth' closes up over the insect's abdomen and hind legs, causing the stamens, which lie beneath the upper petals, to sprinkle pollen over the bumblebee's back. When the insect flies off and then lands on other snapdragon flowers, it deposits its pollen on their pistils, leading to fertilisation. Unlike other species, snapdragons don't have a specific mechanism to prevent pollen from the same flower or the same plant dropping onto the pistil of this plant, because the pistil and stamens are in the same place and both mature at the same time. Despite this, it appears that self-fertilisation simply doesn't happen. Scientists with an interest in plant genetics have been studying snapdragons for many years for this very reason, leading to the discovery of a phenomenon called gametophytic self-incompatibility. This is a fancy way of saying that the flower has the ability to say yes or no, because it can recognise specific genes that express proteins on the surface of the pollen. They say yes when the pollen is from a third-party flower; in this case, the pollen is allowed to grow down through the style, reach the ovary and fertilise the ovule. If the stigma recognises the protein on the surface of the pollen grain as having come from a gene it already has, the answer will be a resounding no: this particular pollen is from the same plant, or worse still, the same flower. The stigma switches to blocking mode and the style resolutely refuses to accept the pollen tube: end of story. No chance of the pollen fertilising this particular flower.

This gene regulation system is so accurate that it guarantees that snapdragon flowers won't be self-fertilised and will therefore retain their genetic variability. And it shows: just look at the range of colours found in wild snapdragon petals.

FOXGLOVE

Fairy fingers

The flower spikes of foxgloves stand proud at the edges of damp woodland on the slopes of hills and mountains in Europe and Central Asia, their long stalks bearing tubular purple flowers speckled with little white flecks.

These tubes are an open invitation to passing bumblebees, calling out to be visited. With their slightly upturned rims, their hidden depths aren't fully visible, but the pathway leading to the base of the flower promises rich pickings. The insect somehow senses that therein lies a source of food – nectar for itself, and both nectar and pollen for its larvae, safely tucked away in their underground colony.

The elongated tubes seem to fit the bumblebee like a glove: just watch them as they are swallowed up and disappear into the depths of the flower! And yet the foxglove seems perfectly designed for human fingers too; hence the scientific name *Digitalis*, from the Latin *digitus* (finger), because the flowers mirror the shape of our fingers. But all is not what it seems: the foxglove might give the illusion of being a cheery, well-behaved plant, but it's still a wild specimen and very toxic to boot. Digitalis leaves and flowers contain large quantities of the molecule digitalin, a powerful toxin that affects cardiac function, even at low doses. However, if digitalin is administered in a controlled manner under medical supervision, it can be an effective way to treat heart problems.

Perhaps the sight of a foxglove sets bumblebees and other pollinating insects' hearts – assuming they have such a thing – racing as they draw close? Not, of course, because they are poisonous, but due to the sheer beauty

SCIENTIFIC NAME
Digitalis purpurea

FAMILY
Plantaginaceae (formerly Scrophulariaceae)

HABITAT
Undergrowth and meadows in upland areas, particularly evergreen forests

WHERE TO SEE
On summer hikes in the Alps, Vosges or Jura mountains, or in alpine gardens like the Lautaret Alpine Garden near Grenoble. It's also frequently cultivated in gardens.

FLOWERING SEASON
June-August

STRATAGEM

The flowers have pollen and nectar guides that show insects the way to the centre of the individual blooms. These are arranged along a long stalk or spike and start to mature from the bottom, male in the first instance, and then female. Insects start to feed from the lower flowers, bringing with them pollen from other foxgloves to deposit on the first female-phase flowers, initiating the cross-fertilisation process.

of the flower, with its little round speckles lining the throat of the tubular corolla. It's a sign, crying out for all to see: 'Free nectar and pollen here: come and get it!'. To some extent, this is misleading advertising: the foxglove hides its pollen – along with the pistil – at the bottom of the tubular flower, sheltered from inclement weather.

Pollen is often what insects are really looking for and that's why the white spots on the lower half of the foxglove's purplish-pink corolla are intended to resemble masses of pollen – to insect eyes at least. The bumblebee is drawn in, firmly believing that the pollen it has spotted is there for the taking – but sadly not, it's a cruel trick. Having come this far, though, the insect might as well go one step further and climb up the petal until it reaches the heart of the flower.

As it continues on its quest for nectar, it inevitably touches the stamens and the pistil, and if it happens to have a few grains of pollen stuck to its furry body, these will come into contact with the stigma and fertilise the flower. Even better if this pollen is from a different flower: bingo! Cross-fertilisation accomplished. Foxgloves rely on an interesting strategy to encourage cross-fertilisation: they exploit the fact that bumblebees tend to feed on the tall flower spike from the bottom upwards. Digitalis flowers also start to open in the same order: the buds are always at the top of the inflorescence. When they first begin to flower, the individual blooms are male: the stamens are mature, but the style isn't – these flowers are therefore halfway up the flower spike. The stamens then wither as soon as the pistil is mature: these flowers are at the bottom of the spike. If a bumblebee always feeds from the bottom to the top, it will begin with the female flowers, then move on to the male flowers where it will end up dusted with pollen. When it lands on a new plant, it once again starts at the bottom and, in so doing, deposits third-party pollen on the female flowers, which means there's no risk of depositing pollen from the same plant because it will only visit the male flowers after offloading its pollen haul.

From top to bottom

Digitalis, Paul Ranson, 1899.

The tall flower spike or inflorescence of a digitalis consists of flowers at different stages of maturity: female flowers at the bottom, followed by male flowers and buds last of all.

This bumblebee is able to land on the corolla and climb inside the flower using the hairs conveniently provided on the petal.

MARTAGON LILY

Moths and butterflies welcome day and night

The lily is a prime example of a symbolic flower: not only for its biblical connotations of the Virgin Mary, but also for its subsequent royal connections (although the origins of the 'fleur-de-lys', so beloved of the French monarchy, are debatable, as this may originally have been an iris or even a gladiolus). Lilies are a frequent sight on florist's stalls with their eye-catching white or coloured petals and long, pollen-coated stamens in bold colours. In springtime, their heady trademark scent fills our gardens and Mother's Day bouquets.

Some wild lily species can still be found in their natural habitats, including the Martagon or Turk's cap lily, native to the ancient forests of mountainous regions. This pink lily has a flower made up of six tepals (a term used to refer to both sepals and petals when they look identical), six long stamens bearing large, dark orange pollen anthers and a long pistil. Its blooms make use of a number of well-known strategies to attract pollinators and promote cross-pollination, as can easily be observed. First of all, the stamens, which release the mature pollen, develop before the pistil; this only matures a few days later, curving through an angle of 90° from its initial straight position, pointing downwards, to ensure that the stigma is directed towards the stamens, which are just beginning to wither.

The Martagon lily, just like other lilies, goes all out on the fragrance front, producing a heady perfume, especially as day turns to night. This is perhaps the first indication that this flower is keen to attract a particular kind of insect – hawk moths – which are out and about at this late hour in early summer. These moths

SCIENTIFIC NAME
Lilium martagon

FAMILY
Liliaceae

HABITAT
Forests and meadows in mountainous areas up to an altitude of 2,800 m

WHERE TO SEE
Many lily varieties are cultivated in gardens, but they are also a stalwart of florist's shops (the Madonna lily, in particular).

FLOWERING SEASON
June–August

STRATAGEM

Lily flowers are easy to spot because of their size and strong fragrance. They have very large stamens and a long pistil, located quite some distance from the nectar at the base of the petals. This enables the stamens or pistil to come into contact with the body of hawk moths, or other moths or butterflies, as they take up the nectar with their long proboscis.

are capable of inserting their long proboscis into any narrow crevice as they hover in the evening air.

When they reach the underside of the flower, they see dark lines or patches showing where to insert their proboscis. The tepals initially provide a visual guide, followed by a different texture, to direct the insect's proboscis to the nectar glands. The bottom of the tepals has a rough texture, apart from in the middle, where the nectar is located, making it easier for the moth to insert its proboscis.

While the hawk moth is enjoying its nectar reward, it hovers sufficiently far from the base of the flower to brush against the long stamens and become dusted with pollen at the same time. Flying on to its next destination, there is a considerable likelihood that the hawk moth will come across another, more mature lily flower. By poking its proboscis into the centre of the flower, just as if it were drinking with a straw, keeping its body away from the tepals, the hawk moth may come into contact with the mature stigma, where it deposits a few grains of pollen clinging to its wings or hairy body, resulting in cross-fertilisation.

Yet the Martagon lily isn't just a flower of the night; it has its fans during the daytime too, although it is quite discerning in its requirements. A bee or a bumblebee would be too small to touch the stamens or the pistil at the same time if they tried to access the nectar. In any case, their tongues aren't long enough to even reach the nectar. But have no fear, the Martagon lily has other visitors in mind: butterflies. And the bigger they are, the better. Unlike hawk moths, butterflies don't hover. As there is quite a large gap between the reproductive organs and the base of the flower, only certain species are able to touch the stamens or the stigma with the tips of their folded wings when they are in the right position to extend their long proboscis and reach the nectar. Large whites and brimstone butterflies are among the few species that have been seen to pull off this remarkable feat.

From top to bottom

These three black-veined white butterflies (*Aporia crataegi*) are extracting nectar with their proboscis as their wings become powdered with pollen.

Botanical plate (detail), 19th century.

Large stamens with straight filaments are ready to deposit their pollen on passing butterflies and moths.

LUPIN

The standard bearer

The flowers of this magnificent garden plant take the form of coloured spikes standing tall above the flower beds in early summer. Yet the ornamental lupin isn't the only species; there are over two hundred different species across the world, some of which are grown as food crops or offer important economic benefits. Consider the white lupin, for example, which produces protein-rich, fleshy pulses that have been eaten in the Mediterranean region since ancient times. Or then again the narrow-leaved lupin (Lupinus angustifolius), widely used for animal fodder in Australia. It can also be used as green manure. Like all plants of the legume (Fabaceae) family, which includes peas, beans, clover, etc., lupin roots play host to bacteria that are able to fix nitrogen from the atmosphere to encourage plant growth. In return, photosynthesis produces sugars that feed the bacteria. As a result, lupins are nitrogen-rich plants and once the plant has been cut back and the foliage left in situ, valuable nutrients are returned to the soil. Adding natural soil improvers dispenses with the need for nitrogen fertilisers, one of the major causes of greenhouse gas emissions from the agricultural sector.

Lupins are also widespread on the plains of North America, forming huge colonies with the ability to transform entire hillsides into spectacularly colourful springtime displays. No wonder the Bluebonnet lupin is the state flower of Texas.

Lupin flowers are typical of plants in the Fabaceae family (formerly Papilionaceae) and use the same pollination process. There's no pollen on view from the outside, nor can you see the stigma, merely a butterfly-shaped flower, arranged symmetrically around its

SCIENTIFIC NAME
Lupinus polyphyllus

FAMILY
Fabaceae

HABITAT
Open, light-filled meadows and upland regions; grows up to 4,800 m above sea level

WHERE TO SEE
Commonly found in cottage gardens, but also widespread on the slopes of Lassen Peak in California.

FLOWERING SEASON
July-October

STRATAGEM

Lupin flowers are tightly packed the length of a tall flowering spike and each individual closed flower resembles a butterfly. When a bumblebee lands on a flower in search of nectar, it presses on the lower petals, which open, revealing the stamens, which in turn place pollen on the insect's underside.

axis. It consists of two parts, which are in fact five petals joined together at their base.

The top of the flower comprises a central petal known as the standard, or banner. There are then two horizontal petals, joined at the base, which close over one another to form the keel. In lupins, this keel is often surrounded by two other petals known as wings. Why not take a closer look at a Fabaceae flower and try and identify the individual parts of the flower? Not only are the flowers similar to butterflies, justifying their inclusion in the former Papilionaceae family, but the terms keel and standard remind us that the flowers are also rather like little boats.

The standard flies high in the middle of the flower and is often a different colour to the wings and keel. Its purpose is to guide the pollinator to the nectar. The wings enhance visibility and attract pollinators, while the keel houses the flower's reproductive organs, well hidden and protected from the elements and herbivorous predators. It also acts as a landing pad and perch for passing insects.

Lupins have a heady scent that complements the undoubted attractions of the flower spike. When they come into range, honeybees, bumblebees and any other pollinating insects are able to distinguish between individual flowers. It then takes a matter of moments for a bumblebee to fly up to the flower and land heavily on the keel so it can dip its tongue into the gap between the keel and the standard to access the nectar. Its rather lumbering body pushes down on the two petals forming the keel, which open up, releasing the pollen-laden flower parts (stamens) and allowing them to touch the insect's underside (or belly). During this process, it's not uncommon for pollen to be transferred to the stigma of the same flower as well, which can lead to self-pollination, a very widespread trait among the Fabaceae family. And yet the lupin's strategy has worked: the bumblebee departs for a new destination, its underside covered with pollen, but sufficiently high up to be unaffected by the insect's frequent grooming activities as it flies. The bumblebee continues its journey, eventually landing on another flower, where it deposits the pollen from the first flower on the stigma of the new flower: cross-pollination in action.

Next page (double spread)
Turkey at large in a field of Bluebonnet lupins and Indian paintbrush in a Texan landscape.

IRIS

A right royal mix-up

Found in many different habitats in a variety of colours and shapes, irises are prized by gardeners and pollinators alike, offering the latter a truly immersive experience. In mid-spring you can find irises on riverbanks, with their feet in water (which helps prevent soil erosion), in gardens across the world, and even on the ridges of thatched roofs, where they help stabilise the roof. The iris has always exerted a certain fascination, ever since Ancient Egyptian times: it was associated with the Egyptian god Horus and cultivated for its cosmetic properties and for the perfume extracted from its rhizomes (underground stems, full of nutrients, which can be used to form new plants). The Romans, in turn, associated the plant with Iris, messenger of the gods, who travelled through the sky at the speed of wind. The varied colours and brief lifespan of iris flowers soon established the flower's links with rainbows. Last but not least, this plant, with its sword-like leaves and proud and stately flowers, has also been linked to the grand royal houses of Europe. It was adopted by the Holy Roman Empire, then became the 'fleur de Louys', the first French king to bear the name Louis. Subsequently, this flower, which features on all royal coats of arms, went on to become the well-known 'fleur-de-lys'.

The iris certainly behaves like royalty; not for it the plebeian ways of other blooms. Its petals assume the form of a standard, not unlike the gondola of an airship, which is rather unusual in itself, but in fact this design is ideally suited to achieving its mission of effective cross-pollination. The large flowers can be seen from a distance and they exude a delicate fragrance that can't

SCIENTIFIC NAME
Iris germanica

FAMILY
Iridaceae

HABITAT
Marshland, wetlands, alongside streams or rivers

WHERE TO SEE
Growing on thatched roofs but also in gardens. Many botanical parks have dedicated iris gardens, including the Parc Floral de la Source near Orléans and Brocéliande Gardens in Brittany.

FLOWERING SEASON
End of May-August

STRATAGEM

The flower conceals its sexual organs beneath its petals, guiding insects towards the nectar in the heart of the flower by forcing them to climb up the three vertical petals, aided and abetted by the beards in the centre. They emerge from the flower sprinkled with pollen. When they enter another flower, the pollen is scraped off on the pistil and cross-fertilisation is the result.

fail to titillate the antennae of passing insects, luring them towards the flower.

But what happens when they reach the flower? The iris has three faces, all identical: long, drooping tepals (this is the term used to describe sepals and petals when they have the same function), and upright tepals that create a kind of ball on top of the flower. There is no visible pollen, although there are clues to its presence. Each lower tepal has central bands, often covered with bristly hairs: this is known as the beard. This little mat forms a perfect landing pad and provides sufficient purchase for insects before disappea-ring beneath the upper sepal. Dare the intrepid insect venture further? Will the sought-after nectar be hidden inside? The insect bravely carries on and receives its just rewards in the form of the hoped-for nectar, hidden away right at the bottom of the flower. However, as it makes its way towards the nectar gland, the roof above its head gradually gets lower and lower until all the little hairs making up the beard force our fuzzy friend to touch the upper wall. Initially, a long, semi-rigid band scrapes along its back, then once the bee has passed this milestone, both its head and back come into contact with the anthers, causing the insect to be showered with pollen. Not that it cares: the nectar is there, a mere tongue's lick away! Having stocked up, the bee turns around, crawls back the way it came, passing beneath the pollen sacs en route, and flies back out of the flower, only to discover another iris just waiting to be visited. Off it goes again: back down the tunnel, squeezing between the beard and the upper tepal, and once more under the rigid band – which is actually a long stigma. As it does so, the pollen from the first flower is transferred to the second flower and cross-fertilisation is the result. The bumblebee continues on its merry way, lapping up the nectar and receiving a healthy dose of pollen, then turns to leave. Interestingly, on the return journey out of the flower, the band of the stigma is more flexible and is no longer receptive, allowing the bee to make its exit unimpeded and, crucially, without losing any of the pollen it has just acquired. Instead, this remains in situ to be scraped off by another stigma concealed deep inside the next iris flower.

From top to bottom

The iris flower keeps its sexual organs well hidden: the stamen is visible beneath the central petal, and attached to this is the pistil, which can't be seen in this photo. The yellow beard isn't a set of stamens; its role is to guide the pollinator towards the heart of the flower.

Irises, Vincent van Gogh, 1889.

Horikiri Iris Garden, Utagawa Hiroshige, 1856-1858.

GIANT WATERLILY
Love boat

Waterlilies, otherwise known as nymphaeas, and lotus flowers are aquatic plant species found throughout the world. They are regarded as highly symbolic flowers in some cultures, the sacred lotus flower of the Buddhist religion being one notable example. Then again, you only have to think of Claude Monet's magnificent waterlilies, painted at his garden in Giverny, or lotus roots, a prized delicacy in Chinese cuisine. However, one of the most impressive specimens of the waterlily family has to be the giant Amazonian waterlily (also known as the Victoria waterlily). In its natural habitat, it grows on backwaters in the Amazon basin. This is by far the largest of all the waterlilies – its leaves can grow to as much as three metres across. These huge leaves, or lily pads as they're known, are supported by large, fleshy ribs, armed with prickly spikes. The rims of the individual pads are upturned with a substantial lip measuring about ten centimetres and the entire surface is coated with water-repellent wax – they almost look like enormous Teflon tart tins! These vast pads can even stand the weight of birds like herons, acting as convenient landing stages for the birds to fish from.

The flowers, though, are just as spectacular: the waterlily's huge blooms can be as much as 40 cm in diameter, although they are short-lived and only last for 48 hours. Nevertheless, this is enough time for them to do what they're there for; they are, in fact, extremely efficient in the pollination stakes.

Once each flower opens, it is female for the first day; only its styles are mature, not the stamens. On the second day, the flower is male: the styles are no longer receptive, but the stamens are now active. The flower

SCIENTIFIC NAME
Victoria cruziana

FAMILY
Nymphaeaceae

HABITAT
Tropical backwaters, swamps

WHERE TO SEE
There are fine examples of this plant at the botanical gardens in Nancy, the Tête d'Or Gardens in Lyon, and in the glasshouses at Chaumont-sur-Loire or Kew Gardens in London.

FLOWERING SEASON
March-July

STRATAGEM

The flower floats on the surface of the water, attached to a stalk that extends right down into the mud, and is made up a great many petals and stamens. These provide a space where beetles become trapped for at least one night, allowing them to mate and feed, but in exchange they pollinate the waterlily.

changes colour to reflect this status too: it is white on day one and pink or magenta on day two.

On day one, the flower gradually opens, petal by petal – there can be as many as sixty in total. By the afternoon, the centre of the flower will have become accessible and the flower, still white at this stage, gives off a fruity aroma which is primarily attractive to beetles. As a further enticement, the middle of the flower is heated: this is due to thermogenesis brought about mainly by cells in the ovaries. This increases the temperature to a constant 10 degrees higher than the outside world – just like a thermostat! Not only does the heat allow scent to disperse more easily, it also provides a cosy environment for local beetles, which crawl between the petals and end up in the central cavity, home to the pistils. As the upper petals close up for the night, folding over each other one by one, they form a roof, effectively trapping the insects inside. Imprisoned in their gilded cage, the beetles aren't too miserable: they have central heating and plenty of food. They may not be able to access the pollen, which is initially protected by membranes, but the plant tissue forming the walls of their snug accommodation is edible, full of sugar and deliciously tender. More to the point, while they're well and truly trapped, the beetles may as well make the most of the opportunity to mate – in a nice, warm environment too. And so these beetles spend their first night of captivity, and well into the evening of their second day, in a snug bolthole, well away from predators, satisfying many of their basic needs.

Twenty-four hours after the beetles arrive, the pistils cease to be receptive, and the stamens mature and become accessible. It may be that much of the pollen ends up inside the beetles, but a fair few grains will also land on their shells. By this stage, the flower will have turned a pink or magenta colour: an external signal that the flower's job is over – fertilisation is complete. Then, on the evening of the second day, the prison finally opens its doors: satiated and pollen-dusted, the beetles are able to stretch their wings before flying off and succumbing to the charms of yet another white flower. Their cargo of pollen grains will inevitably be dislodged onto the stigmas of this new flower, the trapdoors will close and the story will start all over again. And most important of all, the giant waterlily has accomplished its mission: successful cross-pollination.

LOOSESTRIFE
Oil mines

You might come across this plant, with its tall, golden yellow flower spikes rising above glossy green leaves, as you take a walk through marshland, or alongside reed beds. Loosestrife, or lysimachia to give it its Latin name, is commonly found in the marshlands of Europe. Some varieties, like *Lysimachia punctata*, have also been cultivated for use in gardens and on balconies, especially the variegated form with its green and white leaves, or Creeping Jenny, a vigorous plant with the ability to form spreading mats of colour across gardens and roundabouts.

But why have we singled out this particular plant? Well, it appears that the loosestrife is keen to stand out from the crowd: it takes a very different approach to attracting pollinators. Like the vast majority of flowers, the loosestrife's aim is to cross the pollen from one plant with the pistil of a different plant to increase genetic diversity. In order to do this, it uses pollen carriers, pollinating insects in fact, which aren't renowned for providing their services for free. While most flowering plants reward their pollen bearers with sugar-rich nectar or the ubiquitous pollen, which is rich in proteins and lipids, the loosestrife has something different on the menu, something quite unusual, but which turns out to be quite a draw for its cognoscenti. As well as providing standard pollen, lysimachias have opted to lubricate the wheels of their pollination deals: they make oil rather than the classic option of nectar. In the centre of the cup-shaped flowers comprising the yellow corolla with its five petals, insects will come across five pollen-bearing stamens and a central pistil. So far, so normal; but, right at the bottom, where you might expect to find nectar in other flowers, the looses-

SCIENTIFIC NAME
Lysimachia vulgaris

FAMILY
Primulaceae

HABITAT
Wetlands, marshland, alongside streams

WHERE TO SEE
While out walking on the shores of lakes or marshy terrain. Lysimachia punctata is a cultivated species, frequently found in gardens.

FLOWERING SEASON
May-August

STRATAGEM

In addition to pollen, the flowers produce natural oil instead of nectar. This oil is harvested by specific insects, particularly one type of solitary bee which almost exclusively pollinates loosestrife, using the oil to feed its larvae and waterproof its nest to protect against damp.

trife contains a rich kind of oil, which has a number of properties that attract insects.

This oil is particularly fatty and provides an excellent source of nutrition for growing larvae. However, because oil is essentially water-repellent, it never mixes with water and, as such, it also has insulating properties for nests or hives, making it a revolutionary product for housekeeper bees.

The unique properties of the loosestrife's flagship product do not go unnoticed by the compound eyes of pollinating insects. One wild bee species has become such a fan that it is unable to survive without it. Aptly named the yellow loosestrife bee (*Macropis europaea*), this insect has become an expert in collecting the oil: its hind legs have little silk bristles that soak up oil whenever the female of the species takes a dip in lysimachia flowers. As she does so, she also collects a fair bit of pollen. The female of this solitary bee species spends a good part of her life buzzing from flower to flower, dipping in and out of the oil and rubbing her abdomen on the stamens. She has also become extraordinarily adept at thrusting her central legs backwards to repel the unwelcome advances of male loosestrife bees, renowned for their habit of hanging around lysimachia flowers... As she flies from bloom to bloom to collect her provisions, our little solitary bee is in effect dispersing pollen and contributing to cross-fertilisation of the loosestrife plant.

She then takes her spoils back to the nest, which is in fact a little burrow excavated in sand banks less than 15 metres from a patch of loosestrife. Inside the burrow, she will have constructed a central corridor, off which there are two to eight little cells: her offspring's 'bedrooms'. The walls inside each of these cells are freshly coated with oil, and there is a little larder made up of a blend of pollen and loosestrife oil for the larva to feed on over the first two weeks of its life underground before it wraps itself in a cocoon. Ultimately, its metamorphosis into an adult loosestrife bee will complete the cycle the following year, when it emerges just as the loosestrife comes into flower again. In other words, this little wild bee is entirely dependent on loosestrife for its survival. As wetlands start to disappear due to the ever-increasing development of artificial landscapes, this could cause loosestrife to become increasingly rare in the wild, resulting in a worrying decline in its pollination partner.

From top to bottom

Botanical illustration by John Curtis, published in the *Botanical Magazine* in 1822.

A yellow loosestrife bee prepares to top up with oil from a loosestrife flower.

The large, yellow inflorescences of the loosestrife are visible from quite a distance.

N.2295
Pub. by S. Curtis Walworth
Weddell Sc.

EELGRASS

Slippery as an eel

Amateur fishkeepers will undoubtedly have come across eelgrass before, even though they may not have noticed that it has flowers. Why? Because some species of this plant are widely used as the preferred backdrops in freshwater aquariums. In the wild, eelgrass (or *Vallisneria*), and particularly *Vallisneria spiralis*, grows in large colonies of sub-aquatic plants with long strands that can be up to one metre long; they form underwater forests in specific kinds of watercourses. These plants are an extremely important part of the freshwater ecosystem structure, providing a valuable habitat and food for many species of fish and snails. These aquatic plants are not algae, and, like its saltwater cousin (*Zostera*), eelgrass is actually a flowering plant. Algae, although extremely interesting in their own right, do not flower – or at least they didn't the last time I looked!

But flowers need pollinating, I'm sure you must be thinking. With anthers, pollen, styles, stigmas, to say nothing of petals, different colours, scents, warmth and all that jazz? The aim being to exchange genetic material – that's what flowers are for, surely? Transferring sperm cells contained in the pollen from one plant to the ovule of another plant? You know – flowers? To encourage genetic mixing via cross-fertilisation? Isn't that what flowers are all about? Which is fine for plants on dry land, when there are insects flying past. Land-based plants can supply a bit of nectar here and there, or attract pollinators with delightful scents or, at a pinch, make use of the wind. But in water? Really?

This is where it gets interesting: how on earth can you get pollen and pistils to come into contact in a

SCIENTIFIC NAME
Vallisneria spiralis

FAMILY
Hydrocharitaceae

HABITAT
Rivers, watercourses

WHERE TO SEE
These plants can be seen quite readily because they are commonly used in aquariums.

FLOWERING SEASON
June-October

STRATAGEM

Eelgrass is an aquatic plant that uses the surface of the water as a pollination mechanism. It has tiny flowers, some of which are male, while others are female. The male flowers become detached from the plant and float to the surface, where they cluster together to form a kind of raft. Then they come face-to-face with the female flowers, which are held on the surface by a spiral stalk shaped like a spring.

medium that's much more powerful than air? One strategy might be to flood the area with pollen, but this would lead to huge losses; imagine all that pollen dispersed through a three-dimensional environment with underwater currents which, although slower, are much stronger than wind. But wait, there is one place where the laws of physics do work in the plant's favour: on the surface. If the plant can manage to float on the surface, this transportation process will take place in only two dimensions, leading to a much higher chance of successful fertilisation. Another plus point is that the surface tension of water molecules means that anything floating on water has a tendency to gravitate towards other floating items. Water also tends to arrange these items in a certain order; as soon as a slight depression appears on the flat surface, this creates a little slope, which in turn draws in any particles floating on the surface. Eelgrass is clearly well aware of the physical properties of the surface of water and, even better, is able to use these to its advantage to get its flowers pollinated.

It has separate male and female plants: the males produce tiny, lightweight flowers, which detach from the plant and rise to the surface as soon as they reach maturity. Once there, these single flowers open, revealing two pollen-laden stamens. A number of male flowers can be produced at the same time and these tend to cluster together in little rafts thanks to surface tension.

The female flowers, on the other hand, remain attached to long, spiral stalks, which grow upwards, causing the flowers to float just below the surface of the water, creating a depression. With the help of the current, the little rafts of male flowers eventually reach the female flowers, where they are attracted by the depression that the female flower has created on the surface. The pollen comes into contact with the pistil and once the female flower has been fertilised, it retracts and the spirals contract. The seed then ripens underwater, before being released and carried away by the currents, germinating further downstream.

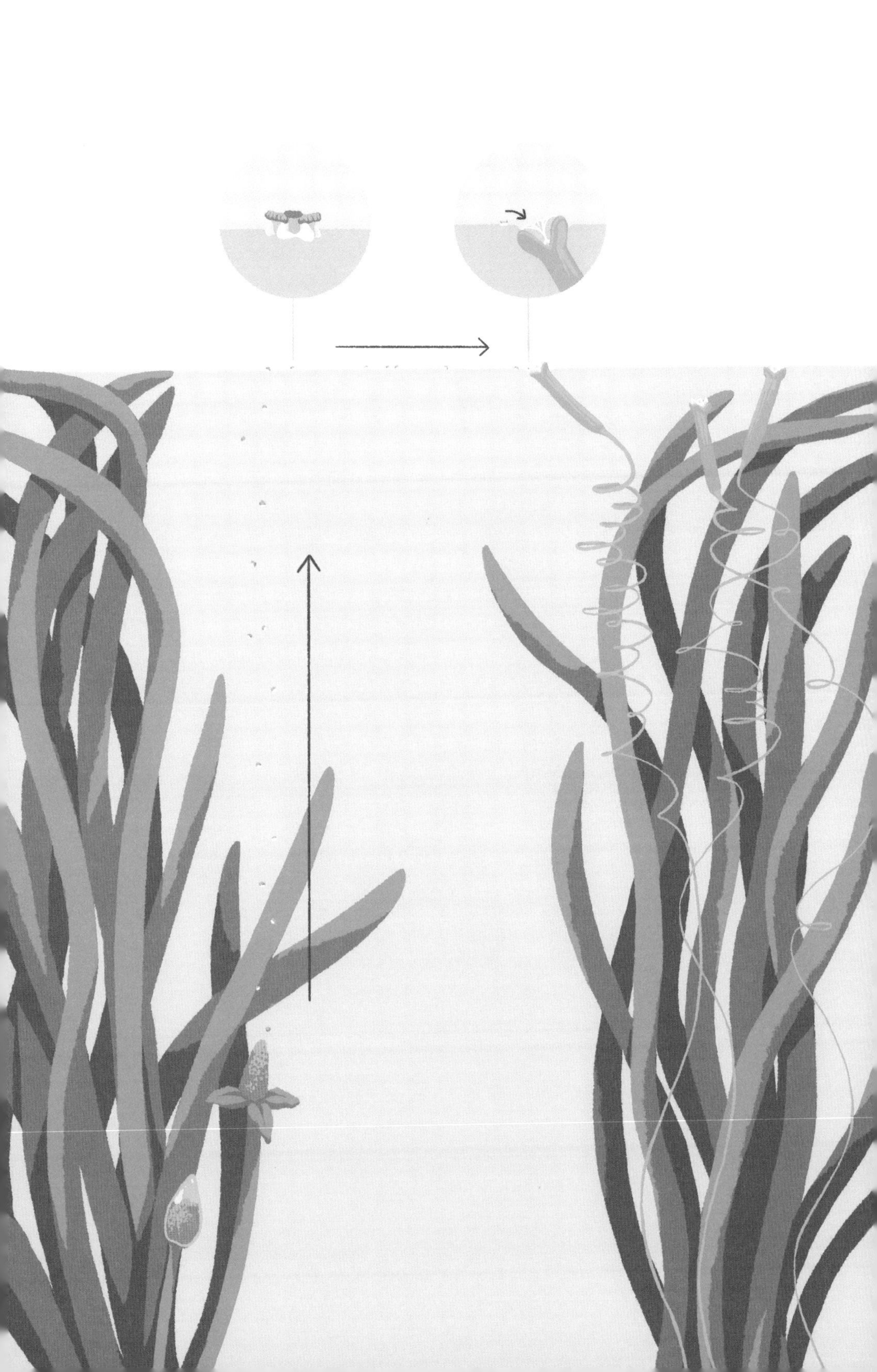

PAPYRUS

Gone with the wind

Often associated with ancient Egypt, the papyrus has inveigled its way into human society as a medium for the first known writings. But have you any idea what kind of pollination strategy it uses?

The papyrus plant grows on riverbanks with its roots in water and its top growth in the air. In its natural habitat, on the banks of the River Nile, for example, strong winds are not uncommon. The alluvial ecosystem (on riverbanks, in other words) tends to be a fairly open environment where wind can blow quite freely. Wind usually follows the direction of the river as it flows from the source to the sea (or vice versa). So why not take advantage of this guaranteed and ever-present mode of transport to disperse pollen to other plants to avoid self-pollination? This is the strategy employed by many riverside herbaceous plants, including reeds and papyrus, as in our case. These species are known as anemophilous plants.

In the wild, papyrus (*Cyperus papyrus*, also known as Nile grass) grows in dense tufts that can be up to five metres tall. Should you happen to be beside the Nile, admiring those fabulous hippos (as you do), you might just espy some papyrus: groves of chunky, green flower heads like giant pompoms at the top of long stems, buffeted by the wind. These rounded tufts are made up of bracts, resembling fine, needle-like leaves, which bear tiny flowers during the flowering season.

However, you might need to know what you're looking for to spot the delicate papyrus flowers themselves. They take the form of little greenish-brown bells, well protected in the centre of the prickly inflorescence with its long bracts, but still accessible to winds from

SCIENTIFIC NAME
Cyperus papyrus

FAMILY
Cyperaceae

HABITAT
Marshlands, watercourses

WHERE TO SEE
Can also be seen growing in gardens. A related species, the umbrella palm or Cyperus alternifolius, is easier to grow and widely available. The Val Rahmeh Botanical Garden in Menton (southeast France) has some beautiful specimens of this plant.

FLOWERING SEASON
June-August

STRATAGEM

The flowers are wind-pollinated – no surprise, as wind is a given alongside watercourses. The individual flowers are tiny, greenish in colour, unremarkable to look at and all at a fixed height on the plant, where they appear in large clusters made up of long, needle-like leaves (or rays). They produce vast amounts of pollen to increase the chances of finding another plant during the dispersal process.

all directions – as they need to be.

What papyrus flowers lack in stature and scent, they certainly make up for in quantity: there's no point in them expending energy to produce artificial adornments like brilliant colours or subtle scents if they can achieve the same aim by producing lots of flowers at minimal cost.

Papyrus flowers are made up of spikelets at the end of a short peduncle, emerging from two tiny, rigid green husks, known as glumes. They are hermaphroditic flowers, having both a pistil and the pollen-bearing stamens, which don't mature at the same time, thus reducing the chances of self-pollination. The stamens hang downwards, swaying in the wind and sending out astronomical clouds of pollen – no doubt contributing to the springtime woes of allergy and hay fever sufferers, along with pollen from conifers and other wind-pollinated trees. These pollen grains are extremely light and can be carried up to several kilometres by the wind before eventually landing on the pistils of mature female flowers on another tuft of papyrus: fertilisation accomplished. Small, dry fruits then form before dropping into the water, floating downstream and finally becoming lodged in a convenient patch of riverbank where they can germinate.

The papyrus – or Nile grass – was regarded as a sacred plant in Lower Egypt and had many uses in Egyptian society: as a writing medium, of course, but also to make baskets and rafts; it was used for culinary purposes and also burned as firewood. The inflorescences were even used during ceremonies to honour the Egyptian gods. Although it has all but disappeared from its original habitat along the River Nile, the papyrus is still an important plant for the Egyptian people. Many other African countries still use it today, to make reed boats in the Niger Delta, for example.

However, the papyrus commonly found growing in flowerbeds in city squares and gardens, or in pots on balconies, is a different species, the umbrella palm (*Cyperus alternifolius*). This is extremely easy to cultivate and also uses a very convenient asexual reproduction mechanism: the tops of the stalks don't have flowers, but, once immersed in water, they form new roots and stems, which spread readily in their natural habitat.

From top to bottom

This large pompom of elongated bracts (a form of leaves) contains miniature papyrus flowers.

Carved stone showing typical birds found near the River Nile and stylised papyrus plants.

Papyrus grows in dense groves along the banks of watercourses.

COURGETTE

Spot the difference

Courgette flowers are well known to foodies, especially devotees of Italian food. They taste delicious, with a slight tang, and are often served deep-fried in tempura batter, sometimes stuffed. Of course, what grows beneath the courgette flowers is extremely good to eat too: the courgettes themselves. These grow all summer long, peeking out from behind the large yellow flowers.

As you stroll through a vegetable garden and come across a courgette plant, you might wonder whether all the flowers conceal a baby courgette behind their jaunty petals. Take a closer look, though, and you'll soon see that only some of the cheery yellow flowers have incipient courgettes at their base. How strange! And should this vegetable garden – or even a balcony – happen to have enough space to grow watermelons, squashes, cucumbers or melons, it's clear they have the self-same big yellow flowers and, guess what, the same thing happens: some flowers have miniature fruit beneath them, while others don't... These species are all related and provide a range of gourds or cucurbits for our delectation; botanists know these as the Cucurbitaceae family.

But wait, what's the story with these cucurbits? Why do they have very similar flowers, yet some turn into fruit and some don't? And what does this mean in terms of pollination? Well, like 5% of all flowering plants, it so happens that, in courgettes, some flowers are male, while others are female. In the cucurbit family, the large yellow flowers have discovered a radical solution to avoid receiving pollen from the same flower, which would lead to self-fertilisation and a potential loss of genetic diversity: both types of flowers, male and

SCIENTIFIC NAME
Cucurbita pepo

FAMILY
Cucurbitaceae

HABITAT
Courgettes and other cucurbits grow on open, sunny sites with ample water

WHERE TO SEE
Often cultivated in vegetable gardens, courgettes can also be found in garden centres. Collections of gourds are also available, notably in the Extraordinary Vegetable Garden of La-Roche-sur-Yon in the Vendée region of France.

FLOWERING SEASON
June-September

STRATAGEM

The flowers are all too visible, being large and yellow. They are either male or female to prevent self-pollination, but they both look as though they have pollen, as the female flowers still have vestigial stamens. This is an excellent way of hoodwinking any insects visiting the female flowers in search of food.

female, grow on the same plant.

Courgette flowers consist of five big yellow petals, all fused together at the base, forming a large goblet opening out into a star. As such, they offer easy access to passing pollinators, with entrances on all sides. The long petals act as roomy landing strips, from where the insects can make their leisurely way into the heart of the flower. They can't fail to notice the characteristic brilliant yellow of the flowers, which, beacon-like, can certainly be seen from afar.

Pollinators are attracted to both types of flower because they both contain nectar – a powerful draw for the insect population. As well as nectar, the male flowers provide pollen as a food source to feed the larvae of all kinds of bees, including bumblebees and wild bees. The female flowers only have nectar on offer – but that's certainly better than nothing. Here though, the problem is that, unlike in hermaphroditic flowers, the lack of pollen may discourage visitors to the female blooms. In other words, the plant needs to find a way of concealing this from prospective pollinators... Hence why, in most cucurbit species, the female flowers have non-functional vestigial stamens. They also have a yellow pistil with a puffy surface that looks not unlike a mass of pollen. This ploy, where the pistil imitates stamens, is very similar to the stratagem used by begonia flowers. Likewise with the male flower: the pollen is provided at the top of the five stamens, all arranged around a fake pistil – once again, this organ isn't functional. All of which means that, even to the compound eyes of a bumblebee, there is no visible difference between male and female cucurbit flowers. In both cases, the bumblebee will assume that both pollen and nectar are available, just as they would be in a standard hermaphroditic flower. However, to prevent the pollen from male blooms on any one plant coming into contact with female blossoms on the same plant, the female flowers on each plant mature later than the male ones. When an insect comes calling, unable to tell male and female flowers apart, it will unknowingly assist with cross-pollination of courgettes, pumpkins, melons and gherkins – all for a nectar reward, of course.

From top to bottom

This female flower has been fertilised and a courgette is starting to form at the base, growing from the ovary.

Bee (*Apis mellifera*) emerging from a male flower.

This male courgette flower is very similar to a female flower.

SAGE
Press here

Prized for its aromatic and medicinal qualities, sage has been grown in gardens since ancient times. Its greenish-grey, downy leaves are the perfect accompaniment to a nice joint of roast pork or as a flavour enhancer for soups. Adding a couple of fresh or dried sage leaves to boiling water also makes for a delicious cup of herbal tea. Both leaves and flowers produce essential oils with a range of medicinal properties: it is these oils that give the herb its unique taste. The same applies to other closely related members of this large family, the *Lamiaceae* (such as thyme, rosemary and mint, for example), which includes many species found on Mediterranean scrubland (or garrigue) and in other fairly arid landscapes. The essential oils they contain, like thymol, are thought to protect the plant at times of drought and also act as a defence mechanism to deter hungry herbivores.

All the flowers in this particular family have very similar morphologies. Even when they're not in flower, it's quite easy to identify Lamiaceae: they usually have square stems.

Sage flowers are a textbook example of the cross-pollination strategy employed by this plant family. Let's take a closer look: a sage flower displays bilateral symmetry; in other words, it doesn't have a circular corolla that visitors can access from any direction. Quite the opposite: in this case, the plant is keen to ensure that visitors arrive from a particular direction – because that's key to what follows. The individual flowers are stacked alternately along tall flowering stems on sizeable shrubs that can grow up to 60 centimetres tall. Each flower is made up of fused purple

SCIENTIFIC NAME
Salvia officinalis

FAMILY
Lamiaceae

HABITAT
Wasteland, fields, scrubland (garrigue)

WHERE TO SEE
Commonly found in vegetable and herb gardens, but also on balconies and even on roundabouts! The botanical gardens in Nice and the gardens at Arzon in Brittany both have a diverse range of sages on display.

FLOWERING SEASON
May-October

STRATAGEM

The flower has tilting stamens: when a bee lands on the lower petal and goes inside, it disappears to harvest the nectar at the base of the flower, activating the stamens, which sprinkle pollen on its back. Then, when it arrives on another flower, the pollen is deposited on the pistil.

petals forming two halves of a gaping mouth: the upper and lower lip, respectively. There is a clear opening between the two lips and the lower lip provides a landing strip from which the pollinator can easily access the nectar at the base of the flower, while still remaining in the same position.

A bee is initially attracted to a sage plant by its fragrance, even before the flower itself comes into view. Only then does it see the flower clusters signalling the presence of food. On arriving at a flower, the bee touches down on the lower lip. The upper petal conceals the reproductive organs (stamens and pistil) to protect them from bad weather. The flower is initially male: only the stamens are mature. Then, when the stamens have run out of pollen, it's the pistil's turn to mature. The stamens are attached to the centre of the flower with a lever-like protrusion at their base. This component, known as the connective, is crucial to the sage's pollination mechanism. As the bee lands on the flower and then shuffles about in search of nectar, it presses the lever, causing the pollen-filled stamens to move downwards and deposit their grains of pollen directly on the insect's fuzzy back. In other words, the stamens are able to move inside the flower, being fixed to the petals at just one point, with the connective, or lever mechanism, acting as a counterweight. When an object – like a bee, or your little finger – touches the connective, the entire stamen tilts downwards. As soon as you remove your finger or the bee flies away, the weight of the connective causes the stamen to tilt back again, returning to its starting position.

Once the flowers are a little older, they are done with the male phase and become female: the anthers are now devoid of pollen and the style is long enough to protrude beyond the upper petal, where it assumes a strategic position. Now visible, it is forked, not unlike a snake's long tongue emerging from the upper petal, ready to snap up any pollen that might drop by.

Let's suppose that our bee, having activated the lever inside a young – and therefore male – flower, then visits an older – hence female – flower. The bee heads right to the base of the flower to find nectar and, in so doing – because older flowers have an elongated pistil that now hangs down inside the centre of the flower – the grains of pollen on the bee's back from the previous flower will come directly into contact with the stigma on the new flower.

The maturity time lag and the ingenious lever mechanism enable the sage to position its pollen in the ideal spot for a bee to carry it safely off to another flower. Even if the bee grooms itself in the meantime, it will be unable to dislodge the pollen on its back and the hidden pollen grains, while not inconveniencing the insect in the slightest, can be transferred smoothly from one bloom to the next. Cross-pollination complete.

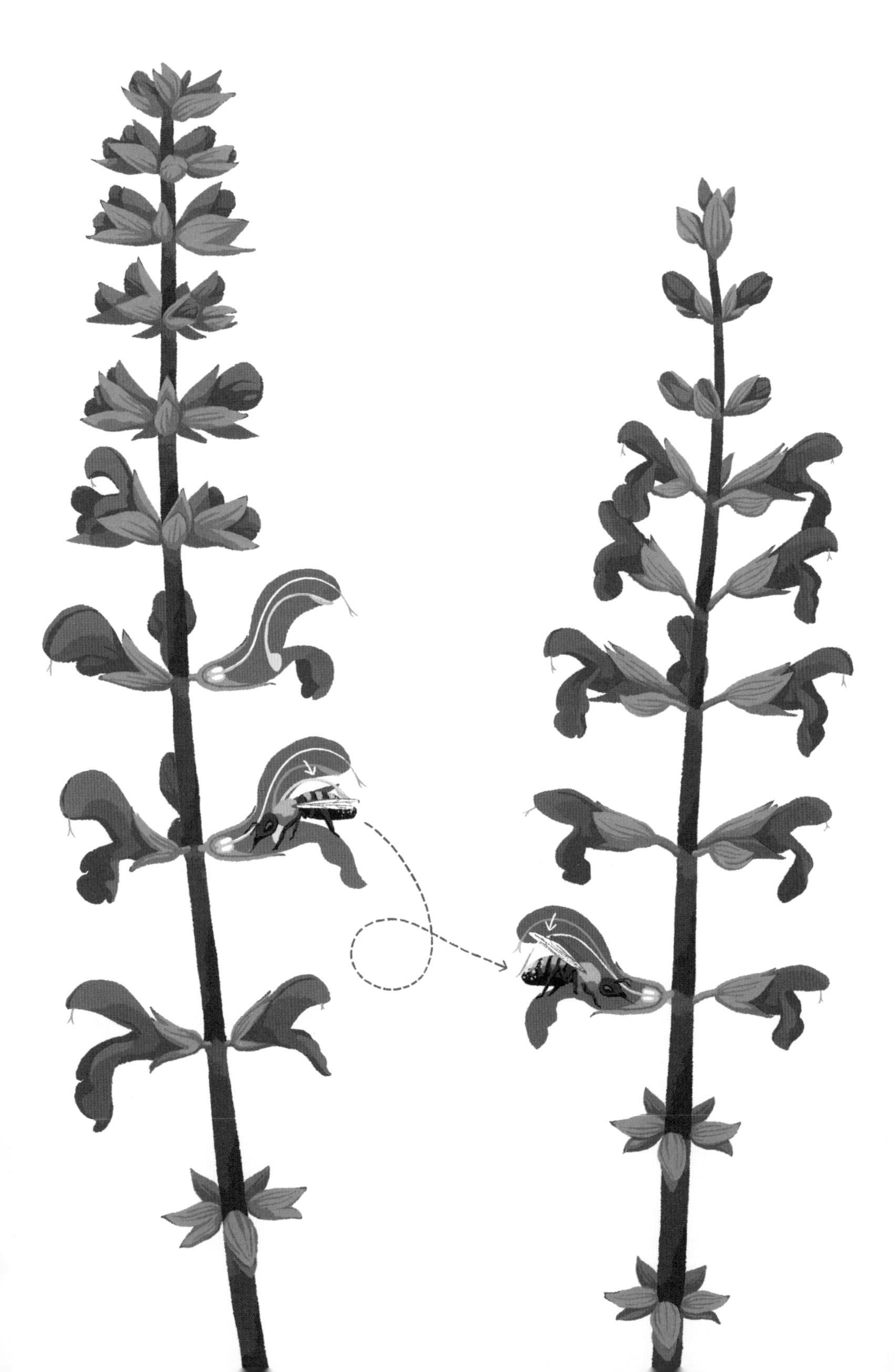

TOMATO

Sensitive to vibration

You'll know, of course, that the tomato is in fact a fruit, not a vegetable – and, like all fruit, it is the result of a flower being pollinated. Pollen from a flower on one plant is deposited on the pistil of a flower on another plant, leading to the simultaneous fertilisation of a number of ovaries, which grow to form tiny seeds as the rest of the flower develops into a fleshy red fruit. The remainder of the flower is clear to see on tomatoes, be they beefsteak or cherry varieties: that little green stalk is none other than the calyx – the collection of sepals or green strips that started life as the flower bud.

Tomato flowers are pretty little yellow stars made up of six pointed petals (the corolla) positioned on six stiff green sepals. A bright yellow cone takes pride of place right in the middle of the star-like petals. The centre of this cone houses the style, topped with the stigma, patiently waiting for pollen to arrive from another flower. A dozen or so pollen tubes are fused together to form a sleeve around the style. Unlike the stamens found in all other flower species, these structures don't leave their pollen exposed to the elements. Oh no – if you want tomato pollen, you have to earn it! This pollen is firmly attached inside these tubes and can only escape through a tiny hole at the top of each tube. Any creature determined enough to extract these pollen grains – to carry them back to its offspring, say – will have to be an acrobat as well as a brilliant dancer, as the tomato doesn't relinquish its pollen lightly. And the conditions have to be just right too: tomatoes, in fact, make their visitors dance – or vibrate – to a very particular tune!

SCIENTIFIC NAME
Solanum lycopersicum

FAMILY
Solanaceae

HABITAT
Open position in full sun with plentiful water

WHERE TO SEE
Tomatoes are queens of the kitchen garden, but can also be grown on balconies. There is a dedicated National Tomato Conservatory in Montlouis-sur-Loire in central France.

FLOWERING SEASON
May-November

STRATAGEM

The flower has tube-like stamens which form a yellow cone in the heart of the petals. The pollen is contained in these tubes, well protected and inaccessible, until a bumblebee comes along in search of pollen and makes the stamens vibrate by causing its entire body to quiver. Only bumblebees and some species of carpenter bees have the ability to pollinate tomato plants.

In other words, the grains of pollen are only released from their tube when the flower vibrates at a specific frequency. Tomatoes are clearly music lovers, as this vibration frequency is nearly the same as the musical note *la*. Indeed, if you were to bring a tuning fork up to a tomato plant, the flower would start to resonate and the vibration would cause pollen to be released.

In the natural world, honeybees, butterflies and beetles are incapable of such musical prowess. This talent requires a combination of flexibility and strength: the ability to vibrate one's whole body at the right frequency to release the pollen while clinging to the tomato flower, upside down.

The bumblebee reigns supreme at this high-flying acrobatic stunt, being capable of what's known as vibratile pollination, or sonication.

Let's not forget that the majority of bumblebee species are social creatures: they live in small colonies where the queen and her sterile daughters go out in search of food, accomplishing pollination in the process, while the males play no part in the collective food-gathering mission. The female worker bumblebees approach a tomato plant, fly up to a flower and clasp their front legs around the pollen cone. They then plunge their long tongue deep inside to access the nectar and start to wriggle. They shimmy and shake, gradually getting faster and faster, until they reach a frequency of 350 Hz, equivalent to the note *fa*, just below the *la* sound produced by the tuning fork. No matter, it's enough – perfect harmony ensues and the tomato flower yields its prize, releasing its grains of pollen. Some of the pollen is mixed with nectar and falls into the little baskets on the bumblebee's hind legs to provide high-protein food for the larvae, waiting back in the nest; the rest gets lodged in the bumblebee's hairy coat. With her mission accomplished, the bee flies away in search of yet another tomato plant, where she repeats her quivering dance, but this time some of the pollen grains are transferred into the tube and fall onto the style, in the middle of the flower, guaranteeing cross-fertilisation.

From top to bottom

The bumblebee clasps the stamens and releases the pollen by vibrating at a high frequency.

Botanical illustration by Giovanni Battista Morandi, published in 1748 in *Hortulus botanicus pictus sive collectio plantarum*.

Once the flower has been pollinated, its ovary swells to form a tomato.

A. Lycoperſicon fructu Ceraſi, luteo. Tourn. Inſt. R. Herb. 150.
B. Lycoperſicon Galeni. Tourn. 150. Solanum Pomiferum, fructu rotundo, ſtriato molli. C.B.P. 167. Mala aurea odore fœtido, quibuſdam Lycoperſicon. J.B. 3. 620. Aurea mala. Dod. p. 458.
A.
B.

PEA

Closed circuit

Take a stroll through a kitchen garden to pick some peas for supper and you're sure to see the beautiful flowers of pea plants – or beans, for that matter; the overall effect is the same. Peas tend to have white flowers, some beans have scarlet blooms, while sweet peas may be pink, but what they have in common is that they all resemble unusual butterflies. They are also very similar to the flowers you'll see on lupins, broom and acacias. All these flowers are members of the Fabaceae family (formerly Papilionaceae). Peas have the floral characteristics of the Fabaceae: they display bilateral symmetry (so-called zygomorphic flowers) and their petals are arranged in a typical structure, halfway between a bird and a boat. Two large white upper petals, veined with green, are fused together to form the upright standard. Then there are two horizontal petals, also white, which bring to mind the hull of a boat; these are known as the wings and they cover two fused green petals forming the keel.

The reproductive organs, in the form of ten pollen-bearing stamens and the central pistil, are protected from above by the wings and from beneath by the keel; they never see the light of day. Peas are far too shy to put their intimate body parts on display for all to see – all that business takes place under cover.

Pea and bean flowers never actually open; they are some of the very rare plants that have opted to rely almost exclusively on self-fertilisation. The closed, scentless blooms of the pea plant are said to be cleistogamous, as can also be the case with some closed flowers in the viola family. However, unlike violets, which only turn to cleistogamy (or self-fertilisation) as a last resort at the end of summer, peas have this as

SCIENTIFIC NAME
Pisum sativum

FAMILY
Fabaceae

HABITAT
Open position in full sun with plentiful water

WHERE TO SEE
In kitchen gardens or fields; this plant has been cultivated since the very beginnings of agriculture in the Middle East.

FLOWERING SEASON
April–July

STRATAGEM

The flowers are butterfly-shaped, their stamens and pistil protected beneath the petals. In most cases, the individual flowers don't open and are fertilised directly using the mature stamens and pistil beneath the petals. Self-fertilisation is their preferred method.

their modus operandi: they keep themselves to themselves, reproducing within each plant, with the whole process taking place concealed behind the wings and the keel.

If we take a peek at what happens beneath these wings as soon as the plants start flowering, the pollen, which is already ripe, is deposited on the stigma, which is surrounded by the anthers – fertilisation: check! This is the method that prevails in over 99% of cases, guaranteeing that the offspring will be identical to the previous generation and adapted to their immediate environment. This preference for self-fertilisation is what led to the pea becoming a domesticated crop. In fact, it was one of the first plants to be cultivated, along with wheat, in what's known as the 'Fertile Crescent', or cradle of civilisation, where agriculture first originated. The present-day pea species is undoubtedly more autogamous (prone to self-fertilisation inside its closed flowers) than the first wild species. When a plant is grown for its nutritional qualities and ease of cultivation, it's hardly surprising that breeders have concentrated on its desired properties to ensure that these are passed on to the plant's descendants. The idea is to select varieties of peas that favour self-fertilisation. However, in the roughly 1% of cases where cross-fertilisation does take place (when a bumblebee manages to force open the fortress created by the wings and the keel, for example), this does guarantee a certain amount of genetic diversity and renewal, helping to perpetuate the species even if their environment changes. It also ensures that adverse mutations within the pea plants do not persist within the species.

This propensity towards self-fertilisation means that the pea is able to create 'inbred strains': these may include pea plants that always have white flowers, or red flowers for that matter, over successive generations. Then again, some varieties are bred for their smooth peas down the generations, whereas others have wrinkled peas. These characteristics were first observed by Gregor Mendel, a monk living in Moravia – now the Czech Republic – back in the 19th century. He was able to demonstrate the principle of genomic inheritance and the phenomenon of dominant and recessive genes in a clear and simple manner.

YUCCA
Pollen offerings

Whether in a pot in an apartment or by the seaside, yuccas are always regarded as exotic in the eyes of the average gardener. This shrub is native to the arid regions of Central America and Southern Europe. It is easily recognised by its tufts of long, sword-like leaves perched on top of slender trunks like miniature palm trees.

While its foliage makes a splendid sight in itself, the plant is even more impressive when in flower. Yucca flowers form large clusters of off-white blooms that spill over the top of the tufts of spiky leaves. These abundant flowers give off a sweet scent as night falls and take on a startling pale glow so they can still be seen in the dark. This is, of course, because they stay open all night, when certain types of insects are active, including a number of different moths. Some of these insect species are particularly fond of yucca flowers, not least the yucca moth (*Tegeticula yuccasella*).

These small moths seek shelter within the yucca flowers, a quiet nook where they can raise their offspring in peace and quiet during the early stages of their life. These insects are truly nocturnal: they spend the day dozing beneath the white bells of the yucca flowers, sheltered from the sun's strong rays. At dusk, the female yucca moth starts to stir. Fluttering from flower to flower, the little moth arrives at a flower in the male phase, with four stamens covered in bright yellow pollen. She takes a substantial amount of pollen in her mouth, wedging it between her large mandibles, and then flies off again, carrying her precious loot through the air.

SCIENTIFIC NAME
Yucca sp.

FAMILY
Asparagaceae

HABITAT
Open, semi-arid or desert areas

WHERE TO SEE
Widely grown as a house plant, it can also be seen on the west coast of France and on the Mediterranean seaboard. There are national yucca collections at the Foncaude Botanical Gardens and the tropical garden at La Londe-les-Maures, both in the South of France.

FLOWERING SEASON
July-August

STRATAGEM

Yucca flowers are pollinated thanks to a symbiotic relationship with a little moth, which transports the yucca's pollen and, in return, makes use of its fruit to feed its larvae. This partnership doesn't work in Europe because the moth doesn't exist here, even though the yucca has since been introduced to these climes. This explains why yuccas don't produce fruit in our part of the world.

The female is on the lookout for somewhere cosy to lay her eggs and the perfect nest for mini yucca moths is in fact an empty ovary inside another yucca flower. So, she visits other flowers: if they are too old, the ovaries aren't quite right, as they are no longer receptive to pollen. She flies on and eventually comes across an ovary that is still mature, even though it's on a fairly young flower. But she still needs to establish whether other yucca moth eggs or larvae have got there first. Using her antennae, our little mother-to-be, still with her stash of pollen clenched between her mandibles, scopes out the odours around the pistil. If she can detect pheromones from other insects of the same species, she will conclude that another insect has beaten her to it and this spot is already taken. She'll just have to try her luck in another flower, a little further away. If this flower turns out to be suitable and the passage is clear, the female busies herself with two jobs that are essential for the survival of her offspring – and therefore of her species – while also ensuring the survival of the yucca population.

She places her ovipositor – a very long, sting-like structure used to penetrate the plant tissue in order to lay her eggs – inside the ovary. She then makes her way down the style, depositing on the surface the generous pollen offering that she has carried all the way from those male flowers she visited earlier. Several of these grains of pollen will start to grow and will in turn fertilise individual ovules. These fertilised ovules, and the development of the associated plant tissue to form the resulting fruit, enable the yucca moth larva to grow. This is to the detriment of the future seed, which forms alongside the egg laid by the female moth, because a gall – a growth on the flower, which helps the insect to develop – forms around the egg itself. However, other seeds will go on to develop. All in all, this is a beneficial partnership for both species – an example of mutualism in action. For once, the pollen isn't devoured by the pollinators, but borne aloft like a precious gift and transported to a female flower. It's almost as if yucca moths understand that cross-fertilisation is the key to seed formation and allows them to feed their larvae at the same time. However, even though the yucca plant has now been introduced to Europe, the yucca moth hasn't followed in its wake and yuccas can only reproduce by vegetative propagation in European countries.

From top to bottom

The white bell-like blooms of the yucca protect the reproductive organs from the rain.

A moth (*Tegeticula maculata*) extracts pollen from the heart of a yucca flower.

Colony of glaucous-leaved yuccas (*Yucca glauca*) in the foreground of the Mitten, a sandstone butte in Monument Valley, Arizona.

HAMMER ORCHID

A striking flower

If you have the opportunity to explore the Australian bush, you're sure to hear a charming story in which the main protagonist is a rather surprising flower. In the dry grass-lands of Southwest Australia, there are some orchids that are well worth a detour. These are the hammer orchids (or *Drakaea*). They aren't much to look at as they aren't brightly coloured, nor do they have a particularly memorable form; they're definitely not the kind of orchid you'd take as a gift for your mother-in-law. Visually, they resemble a blistered, brownish growth, more like a bud or a faded flower than an actual bloom – and yet, the drakaea is at its sexual peak. More on this later, though. First, let's talk about the other protagonist in this story.

This part of Australia has a Mediterranean climate and plays host to thynnid wasps, solitary creatures that lay their eggs in the ground. The females of the species don't have wings, so are flightless, whereas the males, which are larger, are able to fly. During the mating season, the females emerge from the ground and climb to the top of surrounding vegetation. When they reach the top of the stem, they alert the male wasps to their presence by releasing sex pheromones. Lured by this tantalising fragrance, the males swoop down on the females and fly off with them. Their journey can last several hours, during which time the male fertilises the female, but also gives her the opportunity to feed on nectar on a number of flowers before depositing her back on terra firma so she can burrow underground to lay her eggs.

But what about the hammer orchid, I hear you asking. Well, evolution never ceases to amaze: it turns out that, over many thousands of years, some orchid

SCIENTIFIC NAME
Drakaea sp.

FAMILY
Orchidaceae

HABITAT
Open, semi-arid areas with sandy soil

WHERE TO SEE
In its natural habitat in Southwest Australia. This plant is pretty rare and doesn't lend itself to cultivation outside its natural habitat.

FLOWERING SEASON
September-October

STRATAGEM

The flower resembles the body of a female wasp species that lives in the same habitat. The male wasp is taken in by this uncanny resemblance and tries to carry away the part of the flower that looks like the wasp. In so doing, it crashes into the part of the flower containing the reproductive organs and is showered with pollen, which it then transfers to another flower.

species have adapted to take advantage of the strange sexual ballet practised by thynnid wasps.

The drakaea produces a single flower, perched atop a long stem at about the same height as the top of the surrounding grasses, favourite lookout points for female thynnid wasps before they embark on their nuptial flight. The labellum (the characteristic large lower petal seen in orchids) of the hammer orchid takes the form of a black, hairy bulge, slightly sticky and located some distance from the stem, attached to the rest of the flower by a simple joint that allows it to swing. The orchid's sexual organs are at the other end: these include the pollinia (sacs of pollen grains coated in a gluey substance to make them stick to the head of a potential pollinator) and the stigmas, on top of the style, forming the pistil. And that's it. No floral fragrance, as you might expect, merely a blend of chemicals uncannily like the pheromones given off by female thynnid wasps when they are waiting for their male counterparts to come and sweep them off to seventh heaven.

Like the bee orchid, hammer orchids encourage cross-pollination by the phenomenon of pseudocopulation: end of story. The male wasp, tempted by this vision of loveliness and bewitched by the chemical aroma, albeit a pale shadow of the female pheromones themselves, can't help but be drawn in, convinced he's about to meet a fair damsel from his own species. He tries to grab hold of this 'thing', which of course turns out to be nothing other than the labellum of the orchid, still attached to its stem. In his efforts to dislodge this female impersonator, the poor male wasp struggles furiously, causing the plant to perform a seesaw movement: before the wasp knows what's happening, he finds himself striking the orchid's sexual organs, almost like a blacksmith banging his hammer on an anvil. After a few repeat performances of this whole palaver, the pollinia detach from the plant and stick fast to the top of the male insect's head. As soon as the pollinia are released from the plant, the glue surrounding them sets on contact with the air, fixing the two balls of pollen firmly in position on the insect's cranium. By this time the male wasp is tiring and gives up the battle – off he goes in search of another female. He espies another fine specimen, but, yet again, it turns out to be the labellum of a hammer orchid. Same old story – seesaw movement, hammer blows... – but this time the pollen grains come into contact with the stigma and cross-fertilisation is complete. The wasp has served its purpose.

FIG

A flower by any other name

We learn at school that the purpose of any flower is to develop seeds and ultimately fruit; so, theoretically, no fruit without flowers. But where are the flowers on a fig tree?

Perhaps fig flowers are rather shy and retiring? Maybe they've made a conscious decision not to bare all, like so many other plants, exposing their sexual organs to the world at large, putting on tempting displays and giving off sophisticated scents? So it seems – the fig flower looks exactly like a small, underripe fig.

In botany, figs cover many different species from the same family, all of which have this very special kind of flower. They include figs grown in the Mediterranean basin for their sweet fruit, as well as tropical figs, which grow as large trees, widely found in tropical forests; these produce small figs that tend to be enjoyed by monkeys and parrots.

Even though 'modern' domesticated fig varieties, the kind that grow in our kitchen gardens and orchards, are propagated by cuttings and as such have no need of cross-fertilisation, wild or tropical figs have a unique method of reproducing.

The little fig flowers rely on a minuscule wasp (known as the blastophaga) to undertake the task of transporting pollen from one bloom to another – quite unknowingly, of course! Unlike many pollinators, these wasps pay no heed to the visual or olfactory delights of a particular flower: they are far more interested in finding a cosy nest, where the females can lay their eggs in the knowledge that the larvae can develop in peace, well away from predators.

SCIENTIFIC NAME
Ficus carica

FAMILY
Moraceae

HABITAT
Scrubland (garrigue and maquis) around the Mediterranean basin. Some species grow as large trees in tropical forests and mangrove swamps.

WHERE TO SEE
The kinds of fig trees that produce edible fruits are found in kitchen gardens and orchards, especially in the South of France. There are many gardens specialising in figs in that area, with one notable example in the village of Vézénobres. Some tropical fig trees are grown in France as indoor plants: these include fiddle-leaf figs and weeping figs.

FLOWERING SEASON
May-June

STRATAGEM

The fig flower is actually a collection of tiny flowers that have closed in on themselves to form a small green fruit. Pollination is dependent on a symbiotic relationship with a little wasp, which relies on the fig in every stage of its life cycle. It's as though the wasp and some fig varieties have signed an exclusive deal under which the flowers have evolved into their current shape.

And since the nursery provides board and lodging in the form of a roof over their heads and food for the developing offspring, why wouldn't the wasp view this as a paradise on Earth?

It's certainly a cushy number. The fig flower is what's known as a complex flower – similar to the sunflower – where the tiny flowers are well hidden on the inside, closed in on themselves. This is called a syconium, a receptacle containing hundreds, if not thousands, of unisexual flowers (either male or female). This receptacle isn't completely closed; there is a tiny opening (known as the ostiole) at the bottom of the flower, as can be observed on the fruit itself.

Over the course of evolution, a very strange relationship has developed between the blastophaga wasp and the fig, enabling both the insect and the flower to reproduce. In the case of the wasp, this requires a number of stages, covering two types of flowering periods across two different seasons. In winter, the fig tree produces small flowers, each of which contains hundreds of so-called neutral flowers (which are in fact sterile female flowers) and male flowers which form a tightly packed tunnel leading to the ostiole – the exit, in other words. The male figs, or caprifigs, are also known as syconia. Female blastophaga wasps enter the flower through an entrance so narrow that it strips off their wings in the process. They lay their eggs in the neutral flowers, then they die. The eggs develop into larvae and feed on the flowers. After a while, the larvae transform into cocoons, and then into adults: male and female blastophaga wasps. The males emerge first and fertilise the females, after which they die. They will have spent their entire short life inside the inflorescence of the fig. When the females emerge from the inflorescence, they pass through the tunnel of male flowers which have matured in the meantime and are now producing pollen as if there's no tomorrow. These grains of pollen stick to the bodies of the juvenile fig wasps, which then fly off in search of new flowers. From the egg phase, through the larval stage, then onto the males, which are born, fertilise the females and then die, the wasps will have spent a whole season inside the inflorescence and now it's spring again. The fig tree has formed more inflorescences (which also resemble mini figs), but this time these are made up of a combination of neutral flowers and female flowers. The female blastophaga wasps, already dusted with pollen, crawl into these flowers and lay their eggs in the neutral flowers, but as they enter they deposit pollen grains on the female flowers, leading to cross-fertilisation.

MIMULUS
Pollen popper

This flower loves the company of bees or hummingbirds – and it shows! It comes in bright colours that can be seen from far away. Bees or hummingbirds can't fail to notice the unusual structure of this flower with its mismatched lower and upper petals. In fact, it almost looks like a monkey's head, hence the common English name, *monkey flower*. Some of the yellow-flowered species such as *Mimulus guttatus* have little orange spots running down the throat of the flower, pointing the way to the nectar and mimicking grains of pollen at the same time. You may remember that foxglove blooms use a very similar ploy. The entrance to the flower is quite hairy, enabling insects to get a better grip once they reach the lower part of the flower, which also serves as a landing pad.

Hummingbirds and bees delve deep into the tubular part of the bloom to drink their fill of nectar, and, in so doing, the bird's head comes into contact with the pollen-bearing stamens, while bees end up with pollen on their back.

Another species that is almost exclusively pollinated by hummingbirds, *Mimulus aurantiacus*, and widely available in a range of habitats in North America, employs a very effective mechanism to prevent self-fertilisation.

Let's start by looking inside the flower, working from the outside in and down towards the bottom of the tube formed by the corolla (group of petals). We begin at the entrance to the flower, poised like a bee on the doorstep. The female part, or pistil, lies above us. This is white and fused to the roof of the corolla, but, rather unusually, it also ends in a stigma with two open lobes, rather like a mouth or even a waffle iron, ready to snap shut.

SCIENTIFIC NAME
Mimulus sp.

FAMILY
Scrophulariaceae

HABITAT
Open, semi-arid and rocky areas

WHERE TO SEE
A great many ornamental cultivars are available in garden centres.

FLOWERING SEASON
June-September

STRATAGEM

The flower is pollinated by insects or hummingbirds. It has a mouth-shaped pistil which closes suddenly on contact with a pollinator or pollen. If the pollen is from a another flower, the stigma remains closed, but if it's from the same flower, it is rejected by the stigma, which opens up again.

Continuing downwards, we see the stamens, with their load of pollen, and finally, right at the end are the nectaries, source of that precious nectar, so beloved of hummingbirds with their long beaks.

Now you've got the picture, let's go back to the stigma. The two lobes are able to close over each other at the slightest touch from a pollinator. As soon as a foraging hummingbird arrives and comes into contact with the stigma, this snaps shut, like a book, in under five seconds! The stigma remains closed, out of reach of incoming pollen, for almost two and a half hours. So while the hummingbird continues to gather nectar, vibrating as it goes, and setting everything it touches in motion, there is no possibility of self-fertilisation, even if a little bit of pollen falls onto the closed stigma of the flower.

If the hummingbird has visited a different mimulus flower beforehand and is still dusted with pollen, there's a good chance that it will deposit some on the open stigma as it arrives. On contact, the 'jaws' of the stigma will close, remaining tightly shut for several hours – over 24 hours, in fact, if pollen is present. Each grain of pollen will start growing and will continue to do so inside the style for a further 15 hours. Once all the pollen grains have grown and delivered their sperm cells, and a number of ovules have been fertilised, the stigma reopens – but only if there are still some ovules that haven't been fertilised. If all, or almost all, of the ovules have been fertilised, the stigma remains firmly shut. In other words, the mimulus has developed a finely tuned system for controlling the fertilisation process, with the emphasis on just the right amount of pollen to prevent wastage.

Once the stigma has closed and all the ovules are fertilised, the flower is in no hurry to fade. In fact, the fertilised flower may remain in place for days to keep attracting pollinators if the plant still has other flowers to be pollinated.

DAYFLOWER
Pollen - true or false?

This discreet little plant is native to Asia, where it grows away happily in moist soil along the banks of rivers or streams, for example. However, it was introduced to Europe and the United States as a garden plant, and adapted so well to its new environment that it has become an invasive weed in some places. Under its new guise as a weed, this little plant was the first to display resistance to herbicides in Hawaii back in 1957.

Discreet and inoffensive it may seem, but the dayflower is rather good at keeping its secrets. It is a member of the group of flowers that provide only pollen; the dayflower does not produce nectar. Nonetheless, any flower offering pollen to insects shouldn't be surprised if these tasty morsels end up in the stomach of a hymenoptera larva rather than on the pistil of another flower from the same species. Along with others of its ilk that only provide pollen, the dayflower is well aware of the pitfalls. The rose, for example, majors on pollen production: with so many stamens, roses produce masses of pollen on the basis that at least some will end up on the pistil of another rose. The only hitch with this policy is that producing so much pollen requires a lot of effort. And there's no doubt that the dayflower isn't the most diligent of flowers... To get around this problem, it has developed a cunning plan to avoid manufacturing more pollen than absolutely necessary. Instead, the flower takes advantage of its pollinators' appetite for this gold dust by touting its alleged presence, all fresh and tasty, between two blue petals serving as advertising banners. And as we all know, blue and yellow are an eye-catching combination. Unfortunately, those two yellow dots aren't actually

SCIENTIFIC NAME
Commelina communis

FAMILY
Commelinaceae

HABITAT
Understorey of tropical forests, also marshland and paddy fields in Southeast Asia

WHERE TO SEE
It can be found growing in gardens, but has become an invasive weed in the Antilles and on Réunion Island since being introduced there from overseas.

FLOWERING SEASON
June-September

STRATAGEM

The flower has false stamens, which attract pollinators but don't give them any pollen in return. The true stamens, much less showy, sneakily attach pollen to the insect's body, ensuring cross-pollination without any wastage. In other words, the flower is able to save energy by producing minimal amounts of pollen.

generous pollen grains at all; they are merely little growths on the flower that happen to look like pollen!

As soon as pollinators spot the two round blue petals of the flower, they swoop down, eager to feast on what looks like easy pickings of their beloved pollen. But what a disappointment when they land and there's nothing there! In its quest to find at least some pollen, a visiting bee or bumblebee ends up with the real pollen grains stuck to its body. The three fake stamens proudly displayed in front of the petals are merely tantalising lures; three other stamens, fertile in this case, are hidden away beneath the tricksters. Two of these stamens bear dark-coloured pollen and these close on either side of the insect, daubing its flanks with pollen in the process. The third stamen, which is coated in yellow pollen, comes into contact with the insect's abdomen, where it is glued in place. Nor is the dayflower just a pretty face: as well as attaching its surprise packages to the bee in such a way that the insect is unable to consume them (ensuring that they never arrive at their destination), it secretes its pollen in strategic locations on the creature's furry coat. Bees, and particularly bumblebees, tend to be very hung up on hygiene. Their aim is to remove any parasites they might have picked up and eat any random pollen that might have become lodged in their fur. However, some areas, like the insect's abdomen and flanks, are out of reach of their legs, which means that any pollen placed there by the dayflower has a good chance of sticking to their fuzzy body. Duly smeared with pollen, the bee heads off to find another flower, in the hope of getting lucky next time round, but is inevitably drawn to the false stamens of a new bloom. Once the insect's body touches the pistil, pollen from the first flower is transferred and cross-fertilisation is the result.

From top to bottom

The dayflower delicately emerges from its spathe, a form of modified leaf which protects the young flower.

The false yellow pollen, clearly visible in the centre of the flower, attracts pollinators, while the brown pollen gets stuck on the insect's abdomen.

Kingfisher and dayflower, Utagawa Hiroshige, 1830.

ROUSSEA
Lizard juice

The island of Mauritius, in the Indian Ocean, harbours many secrets. As is often the case on these distant isles, cut off from any other civilisations for millions of years, evolution has gone its own sweet way, isolated from the intensive cross-breeding processes observed in continental ecosystems. This has led to some unique back stories and ecosystems with a wealth of species that only grow in this particular place: otherwise known as endemic species.

On Mauritius, wildlife was able to flourish in the absence of any large predators – until humans arrived, that is. No wonder there's an impressive variety of birds and reptiles on the island, including lizards and geckos. There are currently many endemic species that are sadly at risk of becoming extinct, some more so than others. Some have already died out, such as the iconic dodo.

One of these species is the famous blue-tailed day gecko (*Phelsuma cepediana*). This stunning, brightly coloured little gecko loves to feast not only on insects and fruit in its native tropical forest, but also on the spoils of a very specific flower: as produced by a climber that goes by the name of Roussea.

This is a tropical climbing plant with flowers made up of five yellow sepals capped with a circular corolla comprising five amber-coloured petals, fused to form a cup containing ample supplies of nectar.

During the first few days of flowering, only the stamens are mature, but as these wither and fall off, the central style ripens in its turn. This time delay between the initial male phase and the subsequent female phase is the key to the flower avoiding self-fertilisation.

When the gecko arrives to enjoy the strong-scented,

SCIENTIFIC NAME
Roussea simplex
FAMILY
Rousseaceae
HABITAT
Dense tropical forests
WHERE TO SEE
This species is only found on Mauritius and is classified as critically endangered, with fewer than one hundred individual plants at the last count. Attempts have been made to grow this plant at Kew Gardens, in the UK, but without any luck so far.
FLOWERING SEASON
September-January

STRATAGEM

The plant has large flowers with stiff yellow petals. It produces plentiful supplies of viscous nectar, adored by its pollinators, including one particular species of gecko. This is one of the world's very few plants to be pollinated by a lizard. Unfortunately, the nectar also attracts invasive ants and if they have colonised the flower, they bar the gecko from entering.

slightly fermented nectar in a young flower, the top of its head becomes covered with a gelatinous substance secreted by the stamens and containing the pollen grains. The gecko then continues on its way, on the lookout for yet more nectar, and eventually comes across another Roussea flower at a slightly more advanced stage. As it plunges its head inside the flower to reach the precious liquid, the gecko inevitably, and unknowingly, leaves behind a pollen-rich drop of this mucilage, or viscous fluid. And thus ends the story of how the Roussea flower is cross-fertilised!

This whole business is a kind of bargain between two endemic species that have evolved in harmony, left to their own devices for thousands of years. This blooms of this climbing plant are also pollinated by another endemic species, a little bird called the Mauritius grey white-eye. The climber, the bird and the gecko all used to live in perfect harmony, but life isn't all about happy endings and, sure enough, humans came along to throw a spanner in the works, bringing other species in their wake. One day, an invasive ant species started to meddle in the relationship between our three protagonists. These white-footed ants (*Technomyrmex albipes*) try to cover the flowers in clay in which they farm aphids for the sugary digestive juices that they produce. As a result, the ants become very aggressive towards any other creatures that come a little too close to the Roussea flowers, which reduces contact between the climber and the gecko.

Other threats include rats and wild boar – another widespread pest on the island, although bats were originally the only native mammals on Mauritius. Human forestry activities are also contributing to pushing the Roussea towards extinction; so much so that it now features on the International Union for Conservation of Nature (IUCN) Red List of Threatened Species.

British scientists are attempting to grow this climber under glass, notably at Kew, the Royal Botanic Gardens in south-west London. Sadly, seeds that haven't passed through the digestive system of a gecko don't seem to be viable, as they are very sensitive to pathogens. The scientists' latest tack is to germinate the seeds in sterile, pathogen-free environments.

The globalisation of human society – which has turned its back on nature for so many years – is undoubtedly responsible for the sad fact that the Roussea, like many other natural wonders, is now endangered.

BUCKET ORCHID

Perfect for perfume hunters

The orchid family is one of the most populous flowering plant families on Earth – and certainly one of the most diverse. It includes species that employ a range of complicated stratagems to ensure cross-pollination. Orchids are found on every continent on the planet (except Antarctica), and are particularly diverse in tropical regions, where thousands of species grow not only on the ground, but also on trees. They need very little by way of substrate but always reach out to the sun; some are epiphytes, clinging to the branches or bark of trees and tropical climbers.

Bucket orchids are a case in point: originally from Central and South America, these orchids have a very characteristic morphology. They have large sac-like growths, not unlike buckets, formed by their labellum, the modified petal seen in many orchids, which often impersonates the female form to attract male insects. Not just any old males, mind you – and not just any old how! Bucket orchids enjoy a very complex relationship with rather flirtatious male insects, whose aim in life seems to be to please their female counterparts by dousing themselves with high-class scents. These are orchid or euglossine bees – and they have a reputation for stinking to high heaven... Especially the males of the species. These tiny metallic green bees seek to earn their ladies' favours on nuptial flights during which they release blends of fragrances gathered from all across their patch, but especially from the depths of bucket orchids. Eat your heart out, Coco Chanel – male orchid bees, those perfume noses of the tropical forest world, even carry little fragrance bags on the back of their hind legs.

Bucket orchids don't offer rewards by way of nectar

SCIENTIFIC NAME
Coryanthes sp.

FAMILY
Orchidaceae

HABITAT
Dense tropical forests

WHERE TO SEE
If you're lucky, you might see them growing on trees or climbers in the forests of Guyana or Central America.

FLOWERING SEASON
June- November

STRATAGEM

These orchids have flowers with a bucket-shaped petal filled with scented essential oils. Male tropical bees come to forage on the flower to gather these essential oils and use them as a perfume to attract a mate. As they dive into the bowl formed by the flower's unusual shape, they come into contact with the pollinia, which they then transfer to another flower.

or pollen, merely essential oils for these scent-seeking males. The orchid's protuberant labellum forms a bowl, like a pool, to collect the oily liquid secreted by the two glands that are clearly visible at the top of the flower. The male orchid bee draws near, beguiled by the aroma, and promptly starts gathering this elixir straight from the glands themselves. It soon starts to feel rather groggy, drunk on the heady scent, and topples into the pool-like labellum, drenching its wings in the process. Unable to climb out, it ends up immersed in the fragrant oil. After thirty minutes or so, the bee finally manages to clamber up one of the slippery walls of the orchid flower to make its escape, clutching onto hairs as it goes. But this is only the beginning of the male bee's battle: it then has to squeeze through a tiny tunnel and past the pistil, crowned by a stigma. Two pollinia – those compact sacs of pollen grains so often observed in orchids – are then subtly and strategically placed on the wall in exactly the right position to brush against the bee's back. The pollinia detach from the wall and stick onto the orchid bee instead. However, the glue used to fix the pollinia in position doesn't set immediately, so the orchid leaves the bee to stew for a few more moments, just long enough to let the glue set, then finally releases the poor male from its clutches. The design of the flower makes it a struggle for the bee to extricate itself from this narrow tunnel and it takes quite some time.

Even after all its exertions, if the male orchid bee still isn't satisfied with its new scent, it may try to top up the orchid fragrance by leaning a little too enthusiastically into the pool-like labellum of another orchid, kicking the whole saga off again. In it falls, has to fight for its life again to make sure it doesn't drown, and then make its way down that same old exit tunnel, still with sacs of pollen attached to its body – or cuticle, to be precise. As it scrambles out of its unexpected bath, it unwittingly deposits pollen from the previous flower on the stigma, then receives new pollen to take its place. Bravo! The flower's ingenious ploys, combined with its amazing symbiotic relationship with the orchid bee, are responsible for pollination in bucket orchids.

From top to bottom

A male orchid bee carrying pollinia is drawn in by the sweet-smelling droplets produced by this bucket orchid.

The insect has tumbled into the bowl formed by the labellum, which is full of essential oils.

The bee struggles to exit via the tunnel at the end of the bucket orchid's complex flower structure.

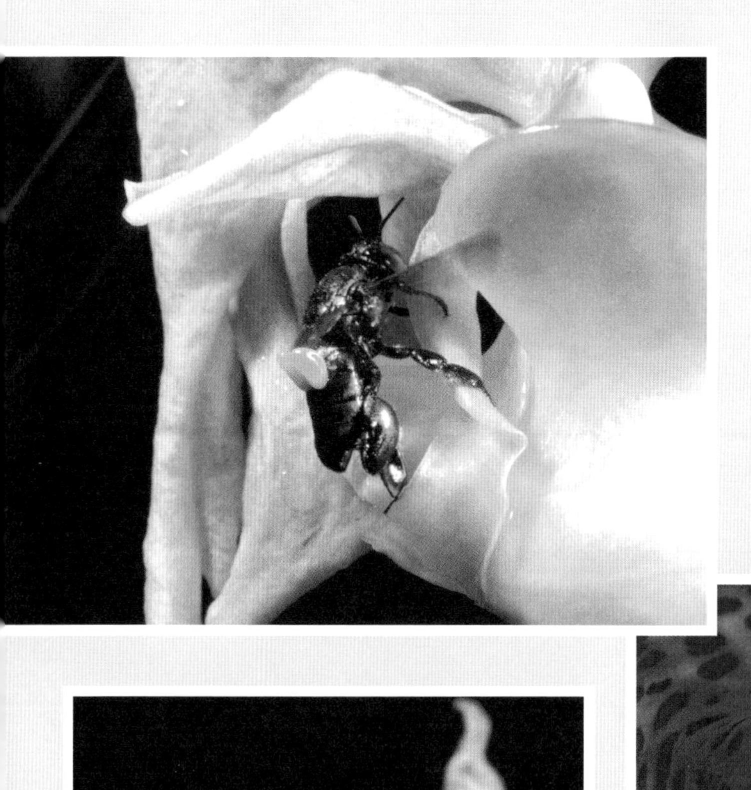

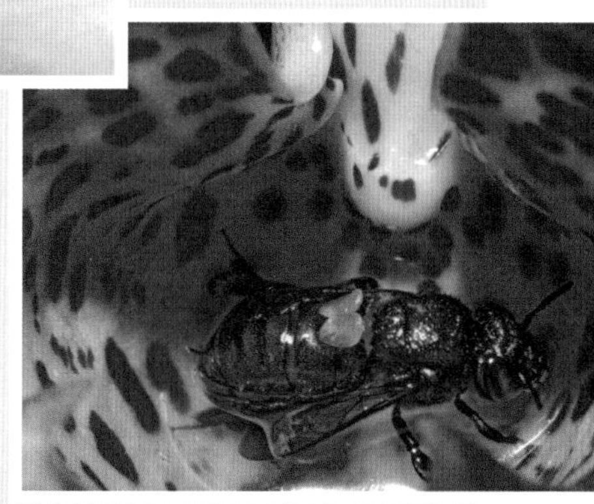

PASSIONFLOWER
A selective show-off

With its ultra-exotic appearance, this climber brightens up the gardens of Europe, especially in the form of one particularly frost-hardy species, *Passiflora caerulea*. Its beautiful circular flowers are quite rare from an anatomical perspective: a sign that the plant is doing everything in its power to attract pollinating bees and bumblebees – and to get them exactly where it wants them!

The passionflower was imported from South America when the Spanish conquistadors made their first voyages of discovery back in the 16th century. Jesuit missionaries were quick to name the flower (just as they lost no time in seizing local treasures), seeing it as a symbol of Christianity that they could use to convert the Pre-Columbian peoples. They believed that the flower symbolised the Passion of Christ, leading to both the plant's scientific and common names. The climber itself, its tendrils and its leaves were thought to resemble the hands and whips used by Christ's tormentors. As for the flower itself, they saw links between the Crucifixion and various parts of the flower: the three black dots on the darker segment are thought to represent the three nails, and the five pollen masses immediately below are the five wounds on Jesus' body. The stunning ring of coloured petals is thought to resemble Christ's crown of thorns, and finally the ten greenish sepals are said to symbolise ten of the twelve apostles. Hence why this flower underpinned the evangelical teachings of the first missionaries, but the Passion to which the flower is so neatly linked might just as well refer to the sufferings the local population would go on to endure under the colonial regime...

In Central and Latin America there are many species

SCIENTIFIC NAME
Passiflora caerulea

FAMILY
Passifloraceae

HABITAT
Forests, hedgerows

WHERE TO SEE
They are naturalised and widely grown in gardens. There is a dedicated conservatory in Saint-Jory in southwestern France. Edible passion fruits can only be grown in the Tropics, however.

FLOWERING SEASON
June-November (until the first frosts)

STRATAGEM

This is an extremely complex flower with narrow, ray-shaped blue petals leading insects directly to the disc of nectar. The flower starts off male, then becomes female to prevent self-pollination. Before the flower becomes female, the pistil is raised and inaccessible to insects. Once the pistil matures, it bends over until it reaches the same level as the stamens.

of passionflower, including the variety that produces passion fruit, or maracuja. In this part of the world, passionflowers are often pollinated by birds such as hummingbirds, or even by bats.

This is partly why the pistil and stamens are fleshy and downturned to make them a little less accessible to the greediest pollinators.

The blue passionflower is now widely available across the globe and has even become an invasive species in places like Spain or New Zealand. Its unusual flower structure helps with the cross-fertilisation process. To start off with, the radial filaments ensure that insects' compound eyes converge on the centre of the flower, where its scent is at its most alluring. As soon as they land on the flower, bees and bumblebees make their way towards the darker centre, where they will find a rill of nectar in the form of a subterranean stream that only the very longest tongues can reach. As they savour their feast, working their way around the central point of the flower, the insects become dusted with pollen as their back touches the mature anthers – at least during the first few days of flowering. The pistil, made up of three segments, each crowned with a little ball, remains upright during this phase, ensuring zero risk of the flower's pollen coming into contact with the stigmas and ruling out self-fertilisation.

On the flower's second day, or thereabouts, the stamens are devoid of pollen and have started to shrivel. The pistil then bends over, reaching the same level as the stamens, which is in turn more or less the same height as a bumblebee or honeybee feeding on the nectar. At this point, if an insect happens to have visited a slightly younger flower just before this one and therefore has pollen on its back, it will brush against the lowered stigmas. Pollen from the first flower will make contact with the stigma, resulting in cross-fertilisation.

The stunning passionflower may have started life as a teaching aid for the evangelical movement, but botanists have long been intrigued by its unique architecture. Now you know its secrets too.

BEGONIA

The art of deception

Frequently grown as indoor plants, begonias are eye-catching specimens, some with white spots on a green background, others with dark pink and green stripes, or perhaps grey marbled effects on frilly red leaves. Begonias are a very diverse plant family, easy to clone by layering, which involves putting a leaf or stem in water for a few days; roots soon start to form, providing a new plant ready to be potted up. However, there are some places where begonias are not just grown in containers, window boxes or flowerbeds. In the wild, begonias are found in meadows, forest margins and many other places where it is hot and humid almost all year round. These include tropical parts of Central and South America, Africa, Asia and Oceania: there are many habitats ideally suited to the world's 2,000 or more begonia species. Haiti is just one example: a number of species grow on this island, some of which were discovered and brought back to Europe by Louis XIV's botanist, Charles Plumier, who is said to have named the species in honour of Michel Bégon, intendant of the former French colony of Saint-Domingue.

In their natural habitat, begonias reproduce in the same way as over 80% of their fellow flowering plants by using animals to carry their pollen from one flower to the next. The begonia is one of just 5% of flowering plants that have different sexes on the same plant, which may have both male and female flowers.

Interestingly, however, begonias are too stingy to provide nectar for their pollinators. Pollen, as a food source, is the only reward on offer to insects.

The anthers, attached to the stamens of the male flowers, carry the pollen. So far, so good: insects are

SCIENTIFIC NAME
Begonia sp.

FAMILY
Begoniaceae

HABITAT
Tropical forests, understorey

WHERE TO SEE
In garden centres, cultivated as indoor plants. The Conservatoire du Bégonia in Rochefort, on France's Atlantic coast, holds the world's largest collection of begonias. It was formerly home to Michel Bégon, after whom the plant is thought to be named.

FLOWERING SEASON
April to October

STRATAGEM

The flowers are either female or male; they don't produce nectar. However, the male flowers do provide pollen to attract pollinators. The female flowers have false stamens that appear to have pollen to make sure that insects still check them out.

attracted to the male flowers to collect pollen and they pick up a smidgen of pollen on their hairs while they're at it. But why should these insects bother to visit female flowers if they don't provide nectar or pollen in return, just their pistil, which has absolutely no nutritional value for a bee?

What if the bee could be easily duped? What if the appearance of the female flowers isn't all it seems? They may not have any pollen to offer, but perhaps it looks as if they have? This is exactly what happens in begonia flowers. The pistil on female flowers looks like a spherical blob of gold, not unlike the stamens on a male flower. Once the bee has gathered pollen from a male flower, it develops an appetite for more. It espies another flower which also looks as though it has pollen to offer, so it plunges headfirst into this bloom in an attempt to devour more pollen – all in vain. Unbeknownst to the bee, what it has managed to do is to deposit pollen on the pistil – which looks like a stamen in this case. In other words, the insect has carried pollen from a male flower to a female flower, leading to the begonia being cross-fertilised.

This deliberate deception could backfire if pollinating insects became wise to what was going on; if they made the same mistake too many times, they might lose patience and decide that begonias aren't worth the trouble, so they might as well go elsewhere to forage. However, it seems that begonias are one step ahead – they produce far more male flowers than female ones. In other words, in the majority of cases, insects are able to find pollen when they land on a begonia bloom. Which means that the whole species benefits from the male/female flower ratio.

From top to bottom

Marmalade hoverfly (*Episyrphus balteatus*) foraging for pollen on a male begonia flower.

Female inflorescence on a begonia.

Begonia, Konan Tanigami, 1917.

BIRD OF PARADISE

The weaver's perch

In European climes, you'll usually see this impressive flower at the florist's; it gives a bold, exotic feel to any floral arrangement. Although it can be grown in hot, humid regions of temperate countries, the bird of paradise flower is most definitely an exotic plant, native to very specific ecosystems in South Africa.

Its scientific name is *Strelitzia* in honour of a British queen, Charlotte of Mecklenburg-Strelitz, wife of King George III. Its common name, bird of paradise, is a nod to its spectacular appearance: it looks remarkably like a fantastic exotic bird with its stiff flower bud (known as a spathe in this instance) protruding like a pointed, upturned beak. At the base of the bud, not unlike the tail and wings of a bird about to take flight, there are three orangey-yellow sepals, which stand out for miles, almost like flags, against the green backdrop. Finally, three blue petals unfurl from the rest, two of these being fused together to form a horizontal projection.

The strelitzia's common name also provides a clue to how the flowers are pollinated; it's an ornithophilous plant – pollinated by birds, in other words. You might think that the flower's resemblance to the actual bird of paradise (in avian form) is a strategy to attract birds on the lookout for a mate, as happens in some orchids, which mimic the appearance of specific insects. But you'd be wrong: the bird of paradise proper does not pollinate the strelitzia. This plant is primarily pollinated by the Cape weaver, a little perching bird (or passerine), usually found in colonies. So, if the flower's eye-catching form isn't a visual lure for the weaver, and it doesn't mimic the bird's shape or colours, what's going on?

SCIENTIFIC NAME
Strelitzia regina

FAMILY
Strelitziaceae

HABITAT
Groves in tropical regions, hedges, tropical forests

WHERE TO SEE
Often seen in florist's shops or garden centres, but it can also grow outside in frost-free areas (Southern Brittany, South of France), such as the Val Rahmeh Botanical Garden in Menton.

FLOWERING SEASON
June-August

STRATAGEM

The plant has large flowers made up of a number of solid petals and sepals. They are pollinated by birds and produce plenty of nectar but have no scent. The arrangement of the petals provides a perch for birds in search of a nectar supply: this perch opens up when the bird touches down, causing pollen to adhere to their feet as they drink.

The striking, contrasting orange and blue colours and the truly impressive stature of the flower are meant to be noticed, guiding birds towards the plant's plentiful nectar supplies. Pollen isn't on the agenda for these birds: it isn't part of their diet. Nor is fragrance part of the bird of paradise's toolkit – like any flower seeking to attract birds, it doesn't waste energy manufacturing aroma compounds, as birds have a very limited sense of smell.

However, in a bid to prevent self-pollination, the strelitzia, like many other flowers, does everything in its power to attach a soupçon of pollen to the bird without it realising to make sure that said pollen ends up on the stigma of another flower. The bird of paradise has all the bases covered: not only does it offer an all-you-can-eat buffet of nectar, it also provides its customers with a convenient perch so they can help themselves in comfort at the optimum angle for accessing the sugary liquid. No wonder the weaver looks so at home on the fused petals forming the bright blue perch. These two fused petals lower down mechanically, releasing the precious nectar at the bottom, previously concealed beneath the third petal. But the flower's feats of engineering don't stop there – this is where the magic happens: the two blue petals making up the perch slowly separate under the bird's weight, automatically revealing a long white strip comprising the mature stamens, which coat the bird's feet and abdomen with pollen. All done and dusted, the weaver continues its quest for nectar. As the bird approaches another flower, the pollen grains attached to its feet are deposited without further ado on the end of the perch, where the stigmas just happen to be located. The new flower receives thc pollen from the previous flower and the plant has achieved its aim.

You can even watch this trick in action yourself: simply press the perch to see the hidden pollen revealed in all its glory.

From top to bottom

A juvenile hummingbird lands on the lower petal of a bird of paradise flower.

Strelitzia flower in full bloom.

Self Portrait with Monkeys, Frida Kahlo, 1943.

Next page (double spread)

The striking colours of bird of paradise petals.

GLOSSARY

Actinomorphic Refers to a flower with a morphology that displays radial symmetry, such as tulips, roses or sunflowers.

Anemophily A form of pollination where pollen is transported from one flower to another by the wind.

Anther Bundle of pollen grains attached to the flower by the filament. The anther and filament together form a stamen.

Autogamy See Self-fertilisation

Bract A leaf at the base of a flower at the intersection between the flower and the stalk. This bract is not always present, or may sometimes take a different form. The bracts of the burdock, for example, are fused together and form the prickly cone that surrounds the inflorescence.

Calyx The sepals of a flower are collectively called the calyx, whether or not they are fused together.

Capitulum (or flower head) Type of inflorescence where a great many flowers, without peduncles, are tightly packed together in a receptacle. These are seen in members of the Asteraceae family like sunflowers, burdocks and edelweiss.

Carpel The female reproductive organ of a flower, consisting of an ovary and its ovule, the style and the stigma. A flower usually has several fused carpels which together form the external female reproductive organs. The carpels collectively form the fruit.

Catkin Flexible, hanging inflorescence as found in nettles or hazelnut trees.

Cleistogamy Self-pollination process used by flowers that do not open, such as peas, or the flower buds of violets.

Cornfield flowers These are flowers that tend to grow in arable fields as a result of soil disturbance.

Corolla The petals of a flower are collectively called the corolla, whether or not they are fused together.

Cross-fertilisation Refers to the fusion of a male gamete (sperm cell) and a female gamete (ovule) from two different flowers, usually borne by two different plants of the same species. This process involves pollen being transported from one flower to another and forms the keystone of the pollination mechanism, which has caused flowers to develop a variety of stratagems to attract pollinators as they have evolved. Cross-fertilisation has the advantage of facilitating genetic mixing between individual plants (two parents with different genes produce the gametes), one of the foundation stones of the evolution of species. This strategy is preferable to self-fertilisation in the vast majority of cases.

Dioecious Refers to plants that have non-hermaphroditic flowers (they have exclusively male and exclusively female flowers) on two separate plants. In other words, some plants are exclusively male and others are exclusively female. Eelgrass is one example of this.

Disc floret This refers to flowers on the flower head of Asteraceae species, where the petals are fused together to form a long tube. These are the flowers found in the middle of sunflowers or burdocks. Each of these is a floret.

Domestication Domestication is an artificial selection process performed by humans to encourage properties that benefit humans. Most vegetables are domesticated; carrots, for example, have been selectively bred from wild carrots over many years.

Entomophily A form of pollination where pollen is transported from one flower to another by insects.

Filament A fine connection between the base of the flower and the pollen grains packed inside the anther. The filament and anther together form a stamen.

Floret See Disc floret

Flower spike Simple inflorescence in the form of a cluster of flowers directly attached to the stem, without a peduncle. This group includes sweetcorn and broadleaf plantains.

Flower stalk The flower stalk refers to the entire stem and the inflorescence carried by the stem.

Gall Tumorous growth on a plant, often caused by an insect laying its eggs or otherwise attacking the plant.

Gamete Male or female sex cell (sperm cell or ovule, respectively). Unlike other plant cells (known as vegetative cells), gametes only have a single copy of the genome of the plant that produced them.

Gametophytic self-incompatibility Gametophytic self-incompatibility is where two gametes (sperm cells and ovules) from the same plant (i.e. sharing the same genetic make-up) do not come into contact with each other, even if the pollen reaches the pistil of the same plant. A very common self-incompatibility mechanism is where the surface of pollen from the same plant is recognised by the surface of the stigma. A chemical interaction takes place, preventing the pollen grain from growing all the way down into the style and coming into contact with the ovule.

Glume Similar to a bract, often found in pairs, located at the base of a collection of flower spikes in a grass like sweetcorn or papyrus.

Hermaphroditic This term describes flowers that have both female (carpels, made up of ovaries and pistils) and male reproductive organs (stamens).

Hymenoptera Refers to an order (or taxon) of insect species with shared characteristics, particularly two pairs of wings that are joined together (creating the illusion that they only have one pair). The term comes from the Latin hymenoptera, which in turn comes from the Greek humenopteros, which means 'with membranous wings'. Hymenoptera include insects that are particularly adept at pollination, such as solitary bees (notably the mason bee) or so-called social bees (such as honeybees or bumblebees), but also cover wasps and ants.

Inflorescence Collection or cluster of flowers arranged on a single flower stalk.

Keel Name given to the two fused petals located on the lower part of flowers from the Fabaceae family, such as lupins.

Labellum Lower petal of orchids, often disproportionately large, which forms one of the plant's attraction ploys, such as when it mimics the body of an insect in the bee or hammer orchid, or when it is filled with essential oils, as in the case of bucket orchids.

Layering Horticultural vegetative propagation technique in which part of the plant (a stem or leaf) is placed in water and then in soil, forming roots and ultimately a new plant. Begonias can be propagated in this way.

Ligule Another word for the ray florets on the flower heads of the Asteraceae family, forming a primary petal. Examples include the outer petals of a sunflower.

Monoecious Refers to plants like begonias that have non-hermaphroditic flowers: they have exclusively male and exclusively female flowers on the same plant.

Nectar Nectar is a sugary solution produced by plants and secreted from the nectaries. It contains varying concentrations of sugars such as sucrose, fructose, saccharose, etc. Nectar may also contain bacteria or yeast which may sometimes ferment and digest the sugars present in the solution. This fermentation process releases heat (as observed in hellebore plants) or alcohol (as seen in the helleborine). Nectar is the main source of energy for pollinating insects like bees, which gather the nectar and store it in their colonies in the form of honey to make sure they have enough food to survive the winter.

Nectary A nectary is a gland that secretes nectar and is often located at the base of a flower, beneath the reproductive organs. Nectaries may also be found at the bottom of growths known as spurs. There may also be false nectaries (or pseudonectaries) that serve to attract pollinators, as seen in love-in-a-mist flowers.

Ostiole A hole at the base of a fig inflorescence (known as a syconium).

Ovary The ovary is the lower part of the female reproductive organ (the carpel). It is located beneath the pistil (comprising the style and the stigma).

Ovipositor This is the reproductive organ of the females of some insect species, used to lay their eggs in fruit, flowers or seeds.

Petal A petal is the main coloured part of the flower. Its purpose is to attract pollinators, but it also serves to protect the reproductive organs, as in campanulas. Petals may be separate (as in poppies) or they may also be fused together to some extent (as in the case of foxgloves). The colours and patterns on petals have a range of different functions associated with reproduction, acting as nectar guides or concentrating heat within a flower. Sometimes, petals and sepals look the same, in which case both are referred to as tepals. Collectively, the petals are known as the corolla.

Petiole Very fine stalk connecting a flower to the main stem. Bracts are usually located at the point where the petiole and the main stem are connected. Not all flowers have petioles: campions and ivy do, whereas plants like sweetcorn, callistemons and plantains do not. The petiole in the female flower of eelgrass is spiral in shape.

Pheromone Chemical substance, similar to hormones, secreted by insects to send chemical signals to each other, whether intentionally or otherwise. For example, females emit sex pheromones to attract males, but there are also defensive pheromones, and some pollinators even leave pheromones behind on flowers like a chemical calling card. Some pheromones can be imitated by flowers as part of their strategies for attracting pollinators, as observed in many orchids.

Pistil The pistil is the upper part of the female reproductive organ. It consists of the style, on top of which is the stigma.

Pollen Pollen is made up of tiny grains (known as pollen grains), which each contain two sperm cells surrounded by thin layers of protective tissue. The layer surrounding a pollen grain is very rigid and made up of various markings in relief. Each flowering plant species produces pollen grains that always have identical markings. Pollen grains are extremely robust and can be found preserved in soil on ancient sites, providing examples of the species growing in previous ecosystems. A grain of pollen can be carried by the wind, animals or water between the male and female parts of a flower from the same species, but on a different plant. This leads to cross-fertilisation. Once this grain of pollen lands on the stigma, it grows down towards the ovule and the sperm cells make contact with the ovule and the carpel.

Pollinia In orchids, these are cohesive masses of pollen grains. There are usually two pollinia, which detach from the flower on contact with an insect, becoming glued to the insect's head or back, enabling them to be transferred to another orchid flower.

Pseudonectary Growth at the base of a flower that resembles a nectar-producing nectary.

Selection Selection relates to a specific union of two individual specimens to give a particular type of offspring that will inherit the characteristics of its parents. Selection may be described as natural: parent plants that are most suited to a particular habitat have the greatest chance of reproducing and will pass on their advantageous traits to their descendants. Over the generations, if the habitat favours a particular trait, natural selection will tend to enforce it. As a result of natural selection, pollination has led to the beneficial evolution of flowering plants and pollinating insects. Humans can also carry out artificial selection when they identify desirable characteristics and fertilise two individual plants accordingly.

Self-fertilisation or autogamy Self-fertilisation refers to the fusion of a male gamete (sperm cell) and a female gamete (ovule) from the same plant (either from the same flower or, as is often the case, two flowers growing on the same plant). Both gametes have virtually identical genes because they come from the same parent. Fertilisation will lead to a fruit and seeds, but these are often much lower quality than in the case of cross-fertilisation. In the long term, a species that relies on self-fertilisation will be less successful in evolutionary terms, especially in an unstable environment, than a species that reproduces by cross-fertilisation. The inbreeding caused by self-fertilisation may also have adverse effects on the fertility of the species. However, some flowers do practise self-fertilisation, including peas, violets, love-in-a-mist, campanulas and tulips.

Sepal A sepal is one of the outermost parts of the flower. Sepals are often green and more rigid than petals; their initial role is to protect the flower, especially when it is in bud. You can see sepals at the base of a tomato, for example, representing the last vestiges of the flower. In some cases, sepals may be coloured and play a role in attracting pollinators, as happens in passionflowers, for instance. Other sepals are coloured and may be confused with the petals, in which case they are called tepals. Collectively, sepals are known as the calyx.

Spadix The spadix is an inflorescence (cluster of flowers), as typically observed in arums. It is the central part, with flowers at the bottom, concealed beneath the spathe, and ending in a pointed tip, as found in arums, usually in the shape of a club-like structure.

Spathe The spathe is the large bract (leaf at the bottom of a flower) surrounding either the flower spike in sweetcorn or the spadix in arums.

Spikelet Collection of flower spikes.

Spur A spur is a very narrow extension of the corolla protruding from the back of the flower. It contains nectar and can only be accessed by the long tongue of a bee or the proboscis of a butterfly.

Stamens Male reproductive organs that

produce and present grains of pollen for the purpose of pollination. A stamen is made up of anthers, which contain not only grains of pollen, but also filaments, by which the anthers are attached to the flower.

Standard The upper petal of a flower from the Fabaceae family.

Stigma The stigma is the upper part of the pistil. Pollen grains come into contact with the female part of the flower on the surface of the pistil.

Style The style is the elongated part of the pistil, connecting the end with the stigma and the base containing the ovary, which in turn contains the ovule. The length of the style may vary – hibiscus plants have a very long style, for example, whereas poppies have very short, swollen styles.

Syconium A particular kind of inflorescence seen on figs, although there is actually no visible flower. It is a receptacle similar to a flower head but is closed in on itself. At the bottom of the syconium, there is an ostiole.

Tassel Cluster of spikelets, as observed in sweetcorn.

Tepal When petals and sepals look the same, they are called tepals.

Umbel An umbel is a particular kind of inflorescence observed in the Apiaceae family (otherwise known as umbellifers), with clusters of tiny flowers on peduncles, forming spherical balls or flat plates, all originating from the same point on the stem. Wild carrots are an example of this kind of flower structure.

Vegetative propagation (**or asexual reproduction**) Reproduction of plants by asexual means. This does not involve pollination, fertilisation or seeds. The offspring is the result of cloning the parent plant. Many plants use vegetative propagation alongside or instead of pollination and fertilisation. These plants often have specific means of propagating themselves, such as the stolons of violets.

Wings The two lateral petals of a flower from the Fabaceae family.

Zoophily A form of pollination where pollen is transported from one flower to another by animals.

Zygomorphy Refers to flowers that have bilateral (or axial) symmetry: the left half of the flower is a mirror image of the right half of the same flower. Examples include snapdragons, sages and pelargoniums.

BIBLIOGRAPHY

Books

Allaby, Michel, *La Scandaleuse Vie sexuelle des plantes*, Paris, Hoëbeke, 2018.

Burnie, David, *Le Mystère des plantes*, Paris, Gallimard, 1989.

Hodge, Geoff, *Practical Botany for Gardeners*, Paris, Marabout, 2014.

Pelt, Jean-Marie, *La Beauté des fleurs et des plantes décoratives*, Paris, Chêne, 2007.

Thomas, Régis, Busti, David, Maillart, Margarethe, *Petite flore de France : Belgique, Luxembourg, Suisse, Paris, Berlin*, 2016.

Willmer, Pat, *Pollination and Floral Ecology*, Princeton, Princeton University Press, 2011.

Scientific articles

Collection, "Les abeilles, familières et extraordinaires", *Espèces. Revue d'histoire naturelle*, no 31, 2019.

Corbera, Jordi, Alvarez-Cros, Carlos and Stefanescu, Constantí, 'Evidence of butterfly wing pollination in the martagon lily *Lilium martagon L.*', *Butlletí de la Institució Catalana d'Història Natural*, 2018.

Fetscher, Elizabeth and Kohn, Joshua, 'Stigma behaviour in *Mimulus Aurantiacus (Scrophulariaceae)*', *American Journal of Botany*, 1999.

Friedman, Jannice, Hart, Katherine S. and Bakker, Meghan C. den, 'Losing one's touch: Evolution of the touch-sensitive stigma in the *Mimulus guttatus species complex*', *American Journal of Botany*, 2017.

Gadoum, Didier, 'La mellite de la lysimaque', *Insectes*, 2009.

Goodwillie, Carol and Weber, Jennifer J., 'The best of both worlds? A review of delayed selfing in flowering plants', *American Journal of Botany*, 2018.

Gottsberger, Gerhard, 'Generalist and specialist pollination in basal angiosperms (ANITA grade, basal monocots, magnoliids, *Chloranthaceae* and *Ceratophyllaceae*): what we know now', *Plant Diversity and Evolution*, 2016.

Liao Hong (ed.), 'The morphology, molecular development and ecological function of pseudonectaries on Nigella damascene Ranunculaceae) petals', *Nature Communication*, 2020.

Wojtaszek, J. W. and Maier, C., 'A microscopic review of the sunflower and honeybee mutualistic relationship', *International Journal of AgriScience*, 2014.

Wróblewska, Anna and Stawiarz, Ernest, 'Flowering of two *Arctium L. species* and their significance as a source of pollen for visiting insects', *Journal of Apicultural Research*, 2012.

Video documentaries

The Private Life of Plants, David Attenborough for the BBC, six 50-minute episodes, 1995.

L'Aventure des plantes, Jean-Marie Pelt and Jean-Pierre Cuny for TF1, 26 episodes, 1982-1987.

Websites

http://www.snv.jussieu.fr/bmedia/Pollinisation/pois.htm

https://www.indefenseofplants.com

ILLUSTRATION AND PHOTO CREDITS

p.7 © Alain KUBACSI / Gamma-Rapho / Getty Images; p.11 © proxyminder / Getty Images; p.12 © Paul Starosta / Getty Images; p.14 © Simon Klein; p.15g © Antony Scott / Getty Images; p.15d © gianni mattonai / Getty Images; p.17 © DEA / ALBERT CEOLAN / Getty Images; p.20 © Sergi Escribano / Getty Images; p.23h © tanjica perovic photography; p.23d © Heritage Images / Getty Images; p.23b © Paul Starosta; p.24 © Sergei Malgavko / Getty; p.28 © VW Pics / Getty Images; p.31h © Pictures from History / Contributor; p.31d © Qi Yang / Getty Images; p.30b © Simon Klein; p.32 © Rehman Asad / Getty Images; p.36 © Pauline Lewis / Getty Images; p.39h © Bridgeman Images; p.39d © Simon Klein; p.39b © Frank Bienewald / Getty Images; p.40 © Focus / Getty Images; p.44 © Mantonature / Getty Images; p.47h © Bildagentur-online / Getty Images; p.47d © Avalon_ Studio / Getty Images; p.47b © David A Johnson / Getty Images; p.48 © REDA&CO; p.52 © mikroman6 / Getty Images; p.55h © Paul Starosta / Getty Images; p.55d © MEPL / Bridgeman Images; p.55b © Heritage Images; p.56 © ullstein bild / Getty Images; p.60 © David Clapp / Getty Images; p.63h © VW Pics / Getty Images; p.63g © Quagga Media / Alamy Banque D'Images / Hemis; p.63d © Arterra / Getty Images; p.64 © blickwinkel / Alamy Banque D'Images / Hemis; pp.68-69 © Jrgen Mu / EyeEm / Getty Images; p.70 © mikroman6; p.72h © RMN-Grand Palais (musée d'Orsay) / Hervé Lewandowski; p.72d © Sergi Escribano / Getty Images; p.73b © Paul Starosta / Getty Images; p.74 © Jacky Parker Photography; p.78 © Arterra / Getty Images; p.81 © picture alliance / Getty Images; p.81d © Paul Starosta / Getty Images; p.81b © ZU_09 / Getty Images; p.82, 85l1 © Arterra / Getty Images; p.85d © Bridgeman Images; p.85b © Bildagentur-online / Getty Images; p.86 © Christopher Ehresmann / 500px / Getty Images; p.89h © DEA / V. GIANNELLA; p.89d © Miroslav Fikar / 500px; p.89b © Johnny Van Haeften Ltd., London / Bridgeman Images; pp.90-91 © Karl Hendon / Getty Images; p.92 © hsvrs / Getty Images; p.96 © Ivana Karic; p.99h © Clive Nichols; p.99d © Claire Plumridge; p.99b © Heritage Images / Getty Images; p.100 © N. Vivienne Shen Photography / Getty Images; p.104 © Abraham Lopez; p.107h © Katharina Kramp / EyeEm; p.107g © Rosemary Calvert; p.107d © Muhammad Owais Khan; p.108 © Frank Sommariva / Getty Images; p.112 © Mantonature / Getty Images; p.115h © DMP1 / Getty Images; p.115d © ZenShui / Michele Constantini / Getty Images; p.115b © Imagebroker / Bridgeman Images; p.116 © Arterra / Getty Images; pp.120-121 © Drew Angerer / AFP; p.122 © sandra standbridge; 125h © Paul Starosta / Getty Images; p.125d © Trudie Davidson / Getty Images; 125b © Florilegius / Bridgeman Images; p.126 © Straublund Photography / Getty Images; p.129h © Sepia Times / Getty Images; 129d © USC Pacific Asia Museum / Getty Images; p.129 © Paul Starosta / Getty Images; p.130 © Christian Peters / EyeEm / Getty Images; p.133h Photo © Whitford & Hughes, London, UK / Bridgeman Images / ADAGP, Paris, 2022; p.133d © imageBROKER/ Andre Skonieczny / Getty Images; p.133b © malerapaso / Getty Images; p.134-135 © DieterMeyrl / Getty Images; p.136 © imageBROKER / hemis.fr; p.140 © John James / Alamy Banque D'Images / Hemis; p.143h © Bridgeman Images; p.143d © Tim Graham / Getty Images; p.143b © georgeclerk / Getty Images; p.144 © Alina Roe / 500px / Getty Images; p.147h CAVALIER Michel / hemis.fr; p.147d © mikroman6; p.147b © Minden / hemis.fr; p.148 © Wolfgang Kaehler / Getty Images; p.152-153 © Terry Eggers / Getty Images; p.154, p.156h © Philippe Gerber / Getty Images; p.157d © Bridgeman Images; p.157b © Heritage Images / Getty Images; p.158 © Wolfgang Kaehler / Getty Images;

p.165h © Florilegius / Bridgeman Images; p.165d © Sabrina Rytz / EyeEm; p.165b © imageBROKER / Helmut Meyer zur Capellen / Getty Images; p.166 © blickwinkel / Alamy Banque D'Images /Hemis; p.170 © Rick Elkins / Getty Images; p.173g © Paul Starosta / Getty Images; p.173d © Boris Guilbert; p.173b © Massimo Piacentino / Alamy Banque D'Images / Hemis; p.174 © Clare Gainey / Alamy Banque D'Images / Hemis; p.177h © Cavan Images / Getty Images; p.177d © Paul Starosta / Getty Images; p.177b © Christian Nze / EyeEm / Getty Images; p.178 © Andrés Pérez Wittmann / Getty Images; p.182 © Paul Starosta / Getty Images; p.185h © picture alliance / Getty Images; p.185g © Ed Reschke / Getty Images; p.185d © Veneranda Biblioteca Ambrosiana/Gianni Cigolini/ Mondadori Portfolio / Bridgeman Images; p.186 © Jenny Dettrick / Getty Images; p.190 DEA / G. SOSIO / Getty Images; p.193h © Ejla /Getty Images; p.193d © Auscape / Getty Images; p.193b © Arterra / Getty Images; 194 © Robert Wyatt / Alamy Banque D'Images / Hemis; p.198 © Rosmarie Wirz / Getty Images; p.202 © Ed Reschke / Getty Images; p.206 © by IAISI / Getty Images; p.209h © Tetsuya Tanooka / Aflo / Getty Images; p.209d © Sepia Times / Getty Images; p.208 © shene / Getty Images; p.210 © Courtesy Anwar Bhai Rumjaun; p.214 © Anna Koldunova / Alamy Banque D'Images / Hemis; p.217 © Oxford Scientific / Getty Images; p.218 © Maarten Van Loon / EyeEm / Getty Images; p.222 © Rizky Panuntun / Getty Images; p.225h © By Eve Livesey / Getty Images; p.225d © I love Photo and Apple / Getty Images; p.225b © Florilegius / Bridgeman Images; p.226 © lathuric / Getty Images; p.229h © Diana Haronis dianasphotoart.com; p.229d © Mike Hill / Getty Images; p.229b Photo © Fine Art Images / Bridgeman Images / Frida Kahlo Foundation / ADAGP Paris, 2022 / Fondation Frida Kahlo; pp.230-231 © eROMAZe

ACKNOWLEDGEMENTS

Thanks to Dr Pierre Tichit and Dr Félix Lallemand for their scientific inspiration – which was beyond inspirational! I'd like to thank Dr Richard Lansdown, a researcher at Kew Gardens, along with his colleagues, for their work to identify eelgrass, Dr Anwar Bhai Rumjaun and Dr Vikash Tatayah, who provided the images of the last Roussea climbers in Mauritius. Thanks also to Dr Mathieu Lihoreau and Professor Andrew Barron who allowed me to plunge headlong into the fascinating world of pollination as part of my doctorate. And special thanks to Loan Nguyen Thanh Lan for her superb illustrations, which combine technical accuracy with magical detail. This book would never have come to fruition had it not been for the efforts of my editor Boris Guilbert, aided and abetted by Ariane: my sincere thanks to them too. And last but not least, to my family, who have always tended their own beautiful gardens and who unwittingly encouraged me to develop my own love of flowers.
To my parents, to Papy, Maguy and Mamie Pitou, and their fabulous vegetable gardens. Thanks to Sara and Adeline for putting up with me during the writing process. Thanks too to Gwen, who accompanied me throughout France and listened so tolerantly to my endless tales...

Published in France by E/P/A

First published in 2023 Great Britain by
Greenfinch
An imprint of Quercus Editions Ltd
Carmelite House
50 Victoria Embankment
London
EC4Y 0DZ

An Hachette UK company

A catalogue record of this book is available from the British Library

Author: Simon Klein
Translator: Claire Cox
Illustation: Loan Nguyen Thanh Lan

HB ISBN: 978 1 52943 021 9
eBook ISBN: 978 1 52943 022 6

Printed and bound in China

10 9 8 7 6 5 4 3 2 1

Papers used by Greenfinch are from well-managed forests and other responsible sources.